THE OMNIVOROUS MIND

THE

OMNIVOROUS MIND

Our Evolving Relationship with Food

JOHN S. ALLEN

HARVARD UNIVERSITY PRESS

CAMBRIDGE, MASSACHUSETTS

LONDON, ENGLAND

2012

All images prepared by Joel Bruss.

LIBRARY OF CONGRESS CATALOGING-IN-PUBLICATION DATA

Allen, John S. (John Scott), 1961–
    The omnivorous mind : our evolving relationship with food /
John S. Allen.
        p.   cm.
    Includes bibliographical references and index.
    ISBN 978-0-674-05572-8 (alk. paper)
    1. Food—Psychological aspects.   2. Food habits—
Psychological aspects.   3. Diet—Psychological aspects.
4. Nutrition—Psychological aspects.   5. Omnivores.
6. Cognition.   7. Brain—Evolution.   8. Human evolution.
I. Title.
    TX357.A453   2012
    616.85'26—dc23   2011047859

*To the memory of my mother, Yaeko Sumi Allen (1928–2006),
an excellent cook in two cultural traditions*

CONTENTS

INTRODUCTION

It is late August, and I have two compelling tasks at hand. First is finishing this book. Second, and in need of even more immediate attention, is reducing the several pounds of zucchini-like tromboncino squash in the kitchen to dice so that it can be turned into a preserved relish. This is my own version of an age-old human dilemma—what to do with an excess of food. Should I share what I can't eat, and with whom? Should I save it for later, and if so, how do I preserve it? Fortunately, I do not have to worry about saving it for another season when food might not be so plentiful; I am preserving it by choice. For our ancestors, however, an excess of food at one time would have been a mixed blessing—not a problem exactly, but a cognitive and social puzzle.

The tromboncino relish is delicious, and I know that months from now, in the middle of winter, it will be a nice reminder of summer.[1] Beyond this, however, I will derive an inordinate amount of pleasure from eating the relish, which is, after all, just a condiment. For me at least, there seems to be a lot of cognitive value in that relish, and this cognitive value is multifaceted. The pleasure of eating it comes from the memory of a summer past, a sense of pride and accomplishment in eating something that is more or less the product of my own hands, and satisfaction from having

direct knowledge of all aspects of its growth, harvest, preparation, and preservation. There is even a sense of security—I might
someday find myself cut off from civilization, but at least I can
survive on my homemade tromboncino relish.

The foregoing paragraphs may simply identify me as just another North American foodie, a member of a culinary subculture
that is the product of a globalized (even as it preaches localism)
food industry that provides unprecedented access to the foodstuffs of the world to those able to afford them. Sure, but that
does not really explain why I feel the way I do about a particular
jar of relish. Food has meaning, it evokes memories, and it shapes
identities. In this book, I will argue that multiple factors—more
specifically, multiple histories—figure into my feeling.

First, there is my cultural history. I'm an American who grew
up in the 1960s and 1970s, and so for me, consuming preserved
and packaged foods was (and is) a daily occurrence. Indeed, it is
very easy for an American, or anyone else living in the developed
world, to survive on a diet consisting of foods of this kind. Now,
of course, these foods generally are not as wholesome as homegrown vegetables, but a willingness to partake of foods that are
far from fresh, and which are in some fashion highly processed,
is a cultural trait common in the developed world. Second, I
come from a family in which food and gardens were fairly important. When possible, my parents grew fresh and seasonal vegetables, although canning the products of their garden was considered
to be a bit over the top and even a little old-fashioned. Ample food
in general was important, and during times when money was
short, it was a point of pride for my parents (who lived through
the Depression and World War II) that there was always enough.
So it is perhaps no surprise that an ultra-homely food, such as
homegrown and home-preserved summer squash, would prompt
a strong sense of memory, pride, and family in me.

Finally, there is the evolutionary history that I share with all other humans. Like other animals, people need food to survive. Natural selection has resulted in behavioral mechanisms that motivate all animals to pursue, acquire, and consume food. Beyond this, in animals with more highly developed cognitive systems, food-related activities may be perceived as pleasurable. In humans, the basic cognitive mechanisms of motivation, pleasure, and reward that we share with other species are all still in place, but they are expressed in and shaped by variable cultural contexts. One of the goals in the study of human evolution and evolutionary psychology is to understand the interrelationship between biology and culture in producing the behaviors, emotions, perceptions, and feelings we see in people today. Consider the pleasure I feel in consuming preserved tromboncino squash months after its harvest: while the feeling is in part the product of my family and cultural environments, it is both deeper and more universal than that. The challenge is not in ascribing a certain portion of that feeling to biology and another portion to culture, but in understanding it as a complex product of these combined histories.

This book is called *The Omnivorous Mind*, and my goal here is to understand how we humans, as a species, use our brains to "think" food. Compared to other animals, we have reached an unparalleled level of cognitive sophistication and intelligence. Our diets are also unique: there are other omnivores, but humans take omnivory to a level beyond simply what is edible. Foods are cultural objects, invested implicitly and explicitly with meaning and significance beyond their nutritive value. We also expand our food universe with cultivation and preparation techniques, some of which may date back millions of years and others that are far more recent.

How we eat and how we think reflect the unique natural history of the human species. On one hand, we are mammals and primates, and to some extent our diets and perspectives on food have been shaped by the millions of years of evolution that we shared with our closest zoological cousins. Over the last five million years or so, however, humans have gone their own way. With increased intelligence have come a more elaborated sense of consciousness, language, increased behavioral plasticity and creativity, and culture. All of these products of the human brain (which itself has changed in structure and grown substantially in size over the past few million years) fundamentally shape our relationships with and perspectives on food and how we go about eating it.

I want to explore these relationships and perspectives on food and eating from a cognitive point of view. Such an approach necessarily must include reference both to our biological, evolutionary past and to the cultural environment in which humans and their ancestors have lived for millennia. By the end of the book, I hope to have made the case not only that food is something that can be understood in terms of the complex cognitive abilities of the human brain, but also that food—its acquisition, preparation, and consumption—has contributed directly to shaping aspects of human cognition in a social and cultural environment. I suggest that we have evolved a "theory of food," a complex cognitive adaptation that each individual uses to organize and engage the food environment in which he or she lives. It is as much a part of the human mind as language, gender, or sociality.

It might be useful to consider language in a little more detail. Language helps define the cultural environment in which all human behavior, including eating, is expressed. It is analogous to a theory of food in that its expression is necessarily shaped by both biology and culture: they are both biocultural phenomena. Steven Pinker describes language as an instinct based, among other things, on the fact that nearly all people, in the absence of a developmental disability or some other condition, and given something resembling a normal developmental environment, learn language.[2] We do not come out of the womb ready to talk, the way a foal is ready to stand and nurse within an hour of being born. But the neural wiring for language is in place at birth, and over the next few years a child becomes an expert in whatever language he or she is exposed to.

Language is as fundamental to the human experience as anything else; it may even be the single critical adaptation that serves to separate us so clearly from our great ape cousins. The language-mediated revolution in cognition and thinking led to the development of complex cultures, the ability to store information collectively, and the possibility of intensive, long-term learning. Although there are connections at all levels between human biology and behavior and the biology and behavior of other animals, everything we do—sex and courtship, violence and aggression, altruism and conciliation, health and illness—is remade by the cultural and cognitive environment we humans live in.

Our diet and approach to food and eating have also been remade by this enriched, language-mediated cultural and cognitive environment. We eat with our brains. Not literally, of course, but for humans, eating is much more than ingestion and digestion. It involves decision making and choice: we do not simply eat what is edible, and we do not always like foods that taste good. Furthermore, food plays a role in our lives that goes beyond simple calories and nutrients.

How do we think food? Like most aspects of our cognition, the process is shaped both by evolved neurological pathways and networks in our brain and by the cultural environment in which we grow up and live. The human brain is the ultimate, evolving source of our culture, but culture in turn shapes the function, and to a lesser extent the structure, of the brain. Our behavior and cognition are truly biocultural phenomena. Consider something as basic as hunger. We have deep-seated brain mechanisms, shared with other mammals, that are important in regulating and monitoring hunger. Yet the feeling of hunger is highly subjective, influenced by individual experiences and mental states as well as by the culture of eating and food preparation.[3]

Like language, eating is a behavior that is integral to the human experience and can fruitfully be explored at many different levels. How is food thought at the different levels at which human cognition is expressed and shaped? Food and foodways can be used to uncover and explore multiple facets of brain function. In the same way that a prism separates white light into its fundamental components, we can use eating behavior as way to uncover and illuminate some of the basic pathways of the working brain. Human diets are cultural phenomena that emerge via the collective activity of individual eaters. Culture is both an external, collective manifestation of brain activity and an amplifier and enhancer of that activity. In order to understand how the omnivorous mind eats, we need to understand the human diet as a phenomenon that is both biological and cultural.

In the rest of this book I will look at the evolutionary, cultural, and neurocognitive underpinnings of the human diet and eating behavior. In Chapter 1, I introduce this approach by examining the widespread appeal of crispy foods and how they might have become so popular. Chapter 2 charts the biological history of the human diet, from its primate origins to the superomnivorous present. I examine the food-related senses in Chapter 3, focusing on how taste can be both biological and cultural. In Chapter 4, I look at the natural tendency to want to eat more, as well as the much less common but no less interesting phenomenon of wanting to eat less. Chapter 5 explores different representations and levels of memory, and how food may be a privileged object of recollection. Classification and categorization are the subjects of Chapter 6. The possible food environment is immense and complex; looking at how humans categorize objects in the world shows how we make sense of our environment by mentally organizing and simplifying it. And in Chapter 7, I look at how the most creative of species exhibits that creativity in relation to food and diet. Chapter 8, the last chapter, elaborates the concept of a theory of food. I hope the reader is prepared to be receptive to this idea, or at least does not find it too much to swallow.

1

CRISPY

The single word "crispy" sells more food than a barrage of adjectives describing the ingredients or cooking techniques. There is something innately appealing about crispy food.

—MARIO BATALI, *The Babbo Cookbook* (Random House, 2002)

WE HAVE ALL AT ONE TIME OR ANOTHER been drawn to the allure of the crispy. Mario Batali runs high-end restaurants featuring wonderful (and often expensive) reimaginings of regional Italian dishes. At restaurants of this kind, the word *crispy* may be a bit too blunt to appear in menu item titles, but it can always be casually mentioned by the server when describing a dish or reciting the specials of the day. We do not go to fast-food restaurants for a personalized, subtle, or sublime dining experience, so there is little cause for restraint in these establishments; *crispy* is freely used as an inducement to buy. When in the early 1970s Kentucky Fried Chicken added a new chicken preparation to their menu, they dubbed it "Extra Crispy." This bit of marketing genius accomplished two things: first, it made it clear that the chicken was not just crispy but *extra* crispy; second, it necessarily reinforced the idea that the "Original Recipe" chicken was itself crispy (any alternative to crispy being unacceptable).

So why do we humans like crispy? The appeal of crispy food appears, like our inalienable rights to life, liberty, and the pursuit of happiness, to be self-evident. Everybody seems to enjoy crispy food. In support of this notion is the fact that crispy foods are very adept at crossing culinary cultural boundaries. A cultural anthropologist colleague of mine used to lament that the evening plane from New Zealand to Samoa always smelled of Kentucky Fried Chicken, as the Samoan passengers made sure to stock up for their families and friends on the way to the airport. Or consider the potato. It did well enough in spreading from the New World to the Old in the preindustrial era, but with the technology that made possible the large-scale production and distribution of crispier forms of the root vegetable (primarily chips and frozen french fries), the potato "came into its own," according to the Food and Agriculture Organization of the United Nations. This organization saw fit to celebrate 2008 as the Year of the Potato.[1] Even in nations where the potato has been supplanted as a staple crop, its availability in crisp and convenient forms helps maintain its overall popularity.

Crispy seems to have the power to penetrate even the most formidable of cultural barriers. For much of its history, Japan deliberately isolated itself from foreign influences; its cuisine is often seen to be the embodiment of this literally and figuratively insular culture. Yet the well-known crispy aspects of classic Japanese cuisine are all adapted from other cultures.[2] Batter-fried tempura was either invented or imported by Spanish and Portuguese missionaries of the fifteenth and sixteenth centuries, who were allowed into the country until Japan severely limited all contact with the outside world beginning in the 1630s. The breaded, deep-fried pork cutlet called tonkatsu is a Japanese adaptation of the schnitzel found in Austria, Germany, and other European countries. Deep frying using only flour or cornstarch as a coating is referred to as *kara-age*, a term that originally meant "Chinese frying" in Japanese. So when you go to a Japanese restaurant and enjoy your very delicious kara-age chicken wings, tonkatsu, and vegetable tempura, you can at least be certain that your California sushi roll appetizer is traditionally Japanese.

Some scientists, such as evolutionary psychologists and bio-cultural anthropologists, become very excited when they see a behavioral or cognitive pattern that seems to transcend cultural boundaries. Quite reasonably, they hypothesize that the pattern may have some underlying biological and evolutionary basis and that it is not solely the product of local environmental or cultural influences. In other words, some patterns and practices appear in different and diverse cultures with such frequency that it is unlikely to be due to convergence or borrowing from another culture. The appeal of crispy appears to be one of these phenomena. The crispy foods themselves may be transmitted from one culture to another, but many cultures seems to be preadapted to receive them with enthusiasm.

Batali's statement at the opening of the chapter frames the hypothesis that crispy foods are innately appealing to humans. At first glance, this seems to be quite reasonable. But *innate* is a strong word—even a fighting word in some social science circles. Like *instinctual*, it conveys the sense that the human brain is hardwired to produce a certain behavior or preference under almost any environmental circumstances. There is broad acceptance that humans possess a language instinct, but can a case be made for an instinct for crispy? Is crispy as deeply rooted in our evolutionary past as language, and is it as culturally transcendent? Words such as *innate* and *instinctual* may be too strong for the appeal of crispy, or maybe we need to adopt a somewhat softer perspective about what these words mean in the context of human behavior and cognition. I look at crispy here as an exemplar of my biocultural approach to the human diet and eating behavior in general. If we want to understand why we like crispy, then we need to understand how we think crispy.

Sources of Crispy: Insects

Where does crispy come from? If we look at the natural world, at foods consumed in their most unprocessed form, sources of crispy are abundant but not terribly appealing, especially to those accustomed to a contemporary Western diet. Insects are probably the crispiest of animal foods thanks to their hard exoskeleton, which is made of a polysaccharide called chitin (though of course insects can also be eaten in their earlier, squishier stages of development, such as grubs).

Insects can be good sources of fat and protein, and throughout the world insects appear as both bit and featured players in human diets. Although Western observers tend to view insects as either a food of desperation or an effete delicacy, the reality in many traditional cuisines is something in between: they are available, so they are eaten.[3] And in many cases, when adult insects with a mature exoskeleton are eaten, they are roasted, grilled, or fried to an extra-crispy state. Here is a nice recipe for grasshoppers from the tribal peoples of Nagaland in far northeastern India:

> Grasshoppers are usually collected after the harvest of the paddy. The wings and stomach of the insect are removed, washed with clean water and then fried in vegetable oil with ingredients like ginger, garlic, chili, salt, onion, fermented

bamboo shoot, etc. Water is usually not added and it is
cooked dry.⁴

That does not sound too bad. Crispy grasshoppers prepared in
this fashion are readily available in the markets of Nagaland and
in other traditional and not-so-traditional markets throughout
the world.

Even to Western observers, the prospect of a nice crispy fried
insect is no doubt more attractive than that of a bug not prepared
to maximize its crunchiness. The widespread consumption of
insects may indeed support the idea that crispy has an innate
appeal. But why do Westerners so definitively reject insects as
food? Anthropologist Marvin Harris considered this issue at
some length.⁵ He argued that Europeans and Americans regard
insects as "dirty and loathsome" *because* they do not eat them, not
the other way around. If insects have no value as food, then their
roles as disease carriers and despoilers of food, as invasive pests,
come to dominate perception of them. But why do insects have
no value as food in some cultures? Harris suggested that if there
are adequate quantities of large vertebrates combined with an absence of reasonable-sized swarming insects, then foraging strategies will not include insects. In other words, we will take meat over
bugs any day. These conditions are met in the northern latitudes
where traditional Western diets originated. However, good-sized
nutritious insects were and are available seasonally in these regions, and other cuisines that originated in these same climates,
such as those of the native North Americans, traditionally made
use of both large vertebrates *and* insects.⁶ Harris argued that the
Euro-American perspective represented an optimal solution to a
specific set of environmental conditions. Although Harris's idea is
interesting, it may have been an overly rational explanation for why

Westerners reject insects as food. As we will see, food choices at both the individual and cultural levels can be influenced by a wide range of factors, and what is and isn't considered to be food is one of the fundamental markers of cultural identity.

Humans are primates, members of the order of mammals that also includes all monkeys and apes and a curious collection of small-bodied forms known as prosimians (lemurs, tarsiers, bush babies, and so on). A quick survey of the diets of primates (see Chapter 2) reveals that many of them eat bugs quite enthusiastically. In fact, the original primates living some 50 million years ago may have been predominantly insect-eaters.[7] Given this insectivorous primate heritage and the fact that the practice of eating insects is quite widespread among humans, there is likely no basis for an innate aversion to eating insects—quite the opposite, in fact. Do we as a species eat insects because many of them are crispy? Or do we like crispy foods because crispy insects were a food of choice among our ancestors? The latter would suggest that the appeal of crispy foods is ancient and cognitively deep-seated. Perhaps there is a connection between crickets and extra-crispy fried chicken, beyond the occasional unwanted visitor to the deep fryer.

Sources of Crispy: Plants

Plants provides us with another source of crispy food au naturel. One of the common associations we make between crispness and plant foods involves freshness. Now, freshness is a multifaceted concept, dependent upon the food itself and the context in which it is obtained, marketed, and consumed.[8] Fresh meat and fish are obviously not very crisp. But crispness and firmness in vegetables (at least those of the leaf and stalk variety) are signs of water retention, and once a vegetable is picked, it not only begins to lose moisture but also undergoes a change in nutrient composition. For example, sugars begin to convert to starches very quickly, as anyone who has compared the flavor of store-bought sweet corn to that just picked from the garden knows. In addition, the nutrients in fresher vegetables tend to be more accessible than those in less-fresh vegetables, especially when the produce is eaten raw. Vegetable foods contaminated by bacteria also tend to lose crispness and gain sliminess.

As historian Susanne Freidberg describes it, the way we eat fresh vegetables today in the developed world is quite unprecedented in human history.[9] Traditionally, anywhere they were consumed, green vegetables were locally produced and seasonal. Today, with refrigeration and industrial production and transport, they can come from virtually anywhere and be eaten at virtually any time. Aggressive marketing emphasizing the healthful benefits of green leafy vegetables helped overcome the perspective that they were secondary foods after cereals and meats. This drove demand, which in turn supported technological advances in production and packaging that led to the development of a "fresher" product, even though it was one that had an origin quite different from that of the traditional hand-harvested, quickly consumed fresh vegetable.

I would argue that this industrialized produce is not really fresh but conveys a sense or facsimile of freshness. Furthermore, the importance of freshness has led to the development and propagation of varieties chosen and bred to signal freshness to consumers at the expense of flavor. In terms of assessing freshness, the "crisp button" in our brains is one that is meant to be pushed. The importance of crispness can be seen in the increased popularity of iceberg lettuce and Red Delicious apples—crisp, visually appealing, but blandly flavored products that have become the culinary poster children for all that is wrong with mass-market produce.

A problem for today's advocates of local food consumption, small-scale production, and organic methods is that the fresh produce we buy at the farmers' market more or less resembles the stuff available at the supermarket. The bell pepper picked on an August morning a few miles outside of town looks a lot like the one grown in a Canadian hothouse in February. As a species, we humans have evolved to assess finished products more than processes. The ability to "read" the signals for freshness, edibility, and palatability has been critical to survival; much less so has been the ability to figure out how the food got the way it is. We also like convenience: for the busy working single mother, the accessibility of food plays an important part in its appeal, much as it did for the Paleolithic hunter. So the value, for both our bodies and the environment, of producing food more sustainably is not always something we take into consideration when we're making our decisions about what to eat.

The contemporary diet in developed countries is often maligned, but anyone wanting to change how contemporary consumers look at the food they eat is going to have an uphill climb. Recent cultural evolution has produced environments that can easily confound the food-related behavior and cognition that evolved over millennia. Contemporary food habits have been shaped over generations by an industrial and technological world of food production and distribution. This industry has become adept at making products that continuously push the food-related evolutionary buttons in our minds. I will talk more about these buttons later.

Despite the marketing that has made salad a year-round part of the contemporary dinner table, of course not everyone likes crisp, raw vegetables. Food writer Jeffrey Steingarten happens to like raw veggies in moderation, but he derides the "salad gluttons" with their "heads bowed, snouts brought close to their plastic wood-grained bowls, crunching and shoveling simultaneously."[10] He points out that many of the leaves, stalks, pods, and beans that we classify as vegetables (a somewhat arbitrary collection of savory plant foods that can include botanical fruits, such as the tomato) typically come well equipped with a range of toxins designed to prevent animals such as ourselves from eating them. In the ancient battle between eaters and the eaten, consumers of plants evolve methods to overcome plants' defenses, while in turn, plants ramp up their defenses or find alternative methods to limit the damage done by their predators. (The same sort of battle goes on between bugs and the eaters of bugs.) One of the alternative methods plants employ is not to fight their would-be consumers but to entice them. Some plants produce sweet, succulent, juicy seed-carrying fruits to attract animals, who then unwittingly become the vehicles of the plants' genetic dispersal by eating the seeds in one location and depositing them in another.

Among our closest primate relatives, we have cousins that are primarily fruit-eaters (frugivores), such as chimpanzees, and those that are primarily leaf- and stalk-eaters (folivores), such as the gorilla. We are more closely related to chimpanzees, and share with them a body size and activity pattern that are more consistent with the roving lifestyle of a creature seeking ripe fruit than with the slow grazing behavior of animals that consume large quantities of high-fiber, low-calorie leaves and stalks. Our ancestral dietary patterns therefore likely tended toward frugivory, which may explain the aversion of some to raw vegetables.

Primatologists have learned, however, that a simple distinction between frugivory and folivory does not always work in the real world. A chimpanzee or a forest monkey may prefer to eat ripe fruit, but such fruit may be lacking seasonally or during a year of drought, or local supplies may be temporarily exhausted. When preferred foods are not available, primates rely on fallback fare— alternatives that are more accessible but less nutritious.[11] For a chimpanzee with no access to ripe fruit or, in some communities, a small monkey to hunt down, fallback foods might include ter- mites or hard nuts, as well as leafy vegetation. The evolutionary significance of fallback foods may equal that of preferred foods. Joanna Lambert suggests that the fact that chimpanzees in the wild often employ tools to obtain some of their fallback foods (sticks to "fish" for termites in their nests or hammerstones to break open nuts) but not their preferred foods may offer insights into the dietary adaptations seen in our earliest ancestors.[12] I will come back to this topic in more detail later in the book.

Most people in the developed world today, with ready access to a full range of food types and products, would probably put in- sects in the category of fallback food, and if they were being hon- est, they probably would throw raw vegetables in there as well. I don't think it is entirely an evolutionary accident that the texture of a fallback food could be a critical part of its appeal for primarily frugivorous primates. The nutritional content of their preferred foods is signaled by the food's taste and its ability to thoroughly sate hunger. Fallback foods, almost by definition, do not possess these qualities, so their appeal must rest on other grounds. And they must have some appeal, because even if they are not first- choice foods, there are times when they are the best choice.

Sources of Crispy: Cooked Foods

Nature provides crispy foods in the form of insects and water- filled, fibrous plant materials. But cooking, a uniquely human technological development, allows us to move beyond this dismal larder. Cooking creates crispy foods that not only have an appealing texture but also often have full and intense flavors. Crispy textures result from browning reactions that occur when foods are heated. One such reaction is caramelization, which causes browning and crisping when sugars are heated to a high temperature. In terms of flavor, the key thing about caramelization is that it converts a single type of molecule (the sugar) into a variety of different molecules that can impart a range of flavor qualities. As Harold McGee writes, "It's a remarkable change, and a fortunate one: it contributes to the pleasures of many candies and sweets."[13]

In terms of our primate heritage, it is clear that we already had a preference for sweet foods before our ancestors invented cooking. Heating sugars provides a whole new, more intense realm of flavor and appeal, but it would not necessarily lead to an expansion of our dietary niche, at least not until industrialized sugar production was established in the mid-nineteenth century.[14] A different set of browning reactions may be more responsible for turning primarily frugivorous humans into true omnivores. The Maillard reaction, named after the French physician and chemist Louis Camille Maillard, who first described it in the 1910s, explains the process of browning and the development of flavor molecules in a wide range of foods.[15] The Maillard reaction begins with a carbohydrate molecule (a free sugar or sugars in starch) and an amino acid (whether free or part of a protein chain). When heated, these molecules form unstable intermediate products that subsequently lead to hundreds of other potential chemical products. Since this process involves both nitrogen and sulfur atoms, derived from the amino acid, a much greater variety of flavors and aromas can be produced compared to the caramelization of sugar.

The Maillard reaction can be used to explain the development of flavors and colors in a huge range of foods. Extreme heat is not even necessary for the reaction to occur, although it speeds up the process greatly. The flavor and brown color of soy sauce, for example, results from the Maillard reaction occurring during the fermentation of steamed soybeans, wheat, and salt.

What does crispiness have to do with the Maillard reaction? Both occur frequently with dry heat cooking methods, such as roasting, grilling, baking, and frying, which are used on meats, flours, and vegetables, and are therefore responsible for many of the qualities we associate with crispiness. When boiled, meat does not reach a high enough temperature for the Maillard reaction to occur at the surface to any degree. However, in dry heat cooking, the meat surface is dried out, which then allows temperatures high enough for the Maillard reaction. This leads to intensified flavors developing in conjunction with the formation of a crispy crust. Similarly, potatoes form a crispy and flavorful surface during deep frying, while the interior can stay moist enough to impede the Maillard reaction as long as cooking times are controlled. As Harold McGee points out, a potato chip is just a french fry that is all surface and no interior.[16] It is therefore all crispy and delicious.

Cooking has been a critical factor in expanding the potential sources of crisp foods in the human diet. When did this expansion begin? Richard Wrangham has been a forceful advocate for the idea that cooking in general has been one of the most important and fundamental technological developments during our evolution.[17] He traces cooking back to long before there were modern humans (*Homo sapiens*, which first began to appear about 200,000 years ago), perhaps as far back as 1.6 million years ago. The kind of people who lived then are called *Homo erectus*. They emerged first in Africa before spreading to much of the Old

World. They had brains that were intermediate in size between those of the living great apes (and our earliest ancestors) and ourselves.[18] They made stone tools, although other evidence of their technology is lacking. The archaeological record generally does not preserve tools made of wood or skins, but early humans almost undoubtedly used them; a few wooden spears from the late *H. erectus* era testify to that fact.

Cooking requires the controlled use of fire. Archaeological evidence of fire can be quite unambiguous if you find clear evidence of some sort of dwelling with a hearth and ash. But what if *H. erectus* was using fire before they began living in any kind of long-term settlement? Wrangham accepts that evidence of human use of fire dating back hundreds of thousands or even more than a million years can be difficult to interpret, but he also points out that even in sites where we know fire was used, the evidence is often lacking. Cooking can be done with small fires that leave not even the slightest hint of their existence on the landscape. Whatever the precise date for the earliest controlled fire, it was certainly at least several hundred thousand years ago, and thus predates the appearance of modern humans. Fire was therefore an important technology before modern humans evolved. As such, it indeed could have been a critical factor in our own evolution.

Cooking is a human universal, in the sense that it is seen in all cultures. Wrangham's model traces how our early human ancestors went from being primarily eaters of raw plant foods, with just a bit of meat, to being consumers of cooked foods involving significant amounts of both plants and animals. Cooking makes a much greater range of plant materials usable to humans as food; the ability to eat tubers, which are starchy and calorie-rich, may have been one of the most significant developments in our evolution. Cooking the flesh of animals, especially the muscle parts, improves its digestibility, and it also can make the tougher parts of the flesh easier to chew. When chimpanzees eat meat, as they do on occasion, they can quickly gobble up the softer tissues, such as the brain, guts, and liver, but they must gnaw away at the muscle for quite a long time. With cooking, our ancestors would have been able to make more efficient use of the whole carcass of larger animals than a chimpanzee (and presumably our earliest ancestors, before cooking developed) ever could. The upshot of all this is that cooking allowed our human ancestors to exploit a greater range of foods, to obtain calories and nutrients in larger packages (i.e., large game animals and big, tough tubers), and to chew and digest them using less energy. These factors all allowed our species to support a large, energy-hungry brain.

I would not want to insist that crispy food was an essential part of the impact of cooking on human evolution; after all, cooking modifies food in ways other than crisping it. Perhaps cooking became established because our ancestors enjoyed the crisping it provided to otherwise soft or chewy foods. It's also possible that the appeal of crispy foods may not have been widespread, but those early human ancestors who enjoyed them were more likely to keep on cooking, thus accruing long-term benefits over evolutionary time. However, the benefits of cooking and the influence those may have had in the development of our species may provide at least part of an explanation for why we so enjoy crispy foods. Cooking serves as the earliest example of how technology can build upon, modify, and amplify a preexisting dietary preference. Our preference for crispy may have originated with insects and fallback plant foods, but cooking made different foods crispy and moved that preference into the center of our diet. Today, industrialized cooking has made crispy foods available on an unprecedented scale in the developed world. As many are aware, it is very easy to eat too much crispy food of the modern variety. An off button to stop the eating of crispy food, as well as other foods that are innately appealing, did not seem to evolve along with the on button.

The Chewing Brain

We experience crispy by putting food in our mouths and chewing it. We chew with our teeth, which are embedded in both the upper jaw (maxilla) and the lower jaw (mandible). Four pairs of muscles, known as the muscles of mastication, anchored on the cranium and extending to the mandible, work in concert to move the mandible.[19] Babies are not born with very well-developed muscles of mastication, and they also have to learn to control them. Newborn babies actually suckle using some of the small muscles of facial expression rather than the muscles of mastication (for which mothers are probably thankful).

One thing that strongly distinguishes modern humans from even our closest evolutionary relatives is the small size of our jaws. Although we might find the shape of our skulls pleasing, if we compare our skulls to those of the great apes and other members of our zoological family, ours are oddly misshapen, with an enlarged braincase (cranium) sitting over a ridiculously flat face and small jaws. This reduction in the size of the chewing apparatus, which was by no means a requirement for having a large brain, suggests to most researchers that there was a critical shift from relying on our mouths and hands to process our foods to using stone tools, cooking, and other technologies. In fact, a reduction in the size of the teeth (although not always also in the size of the jaw) is a general characteristic that serves to separate members of the genus *Homo* from other apes who walked on two legs, both at present and when the first members of *Homo* began to appear some 2 million years ago (more on this in Chapter 2).

So we all have little teeth and little mouths, but how do we control what we do with them? This gets us to the brain, but before we start looking at the role the brain plays in how we chew, it might be useful to review some brain basics. (Warning: quite a bit of information is coming in a short space.)[20] The central nervous system consists of two main parts, the brain and the spinal cord. The latter consists of a thick bundle of fibers that runs through the vertebrae of the back, connecting to the nerves of the body (peripheral nervous system) and merging into the brain by transiting through a large hole at the base of the skull, the foramen magnum.

The brain itself sits protected within the cranium. It has three major parts: the brain stem, the cerebellum, and the cerebrum. The brain stem connects directly to the spinal cord; in addition, most of the nerves of the head and jaws (the cranial nerves) enter the brain through the brain stem. Among other functions, the brain stem helps to regulate complex motor patterns and breathing, and it is also critical in the regulation of sleep and consciousness. The cerebellum, or "little brain," sits tucked under the larger cerebrum. It is important in the control of voluntary movements, as well as balance and posture.

The tissue of the brain has traditionally been divided into two types: gray matter and white matter. The gray matter is the part of the brain that is made up largely of nerve cells, or neurons. These neurons tend to be located in the wrinkled outer surface of the cerebrum, called the cerebral cortex. The cerebral cortex has both grooves, known as sulci (sing. *sulcus*), and ridges, called gyri (sing. *gyrus*). The folding of the cortex allows for an increased number of neurons to be packed into a given volume.

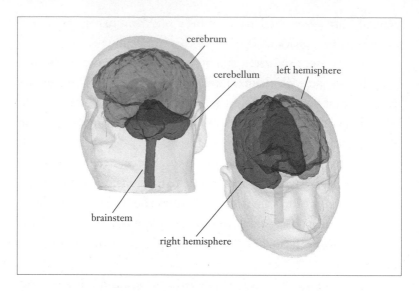

The major subdivisions of the human brain.

The cerebellum is even more tightly wrinkled than the cerebrum, and the neuron density there is higher than in the cerebrum. (Another contributing factor to neuron density, besides the folds, is that one of the cerebellar neuron types is quite small.)

Neurons also aggregate in clumps or clusters (nuclei) that are typically embedded in the interior of the white matter. The white matter consists of non-neuron cells and the long processes (extensions) of neurons through which they communicate. There are two types of neuron processes: the axons are used to send signals from one neuron to another, while the dendrites have sites at which messages from axons can be received. These neuronal processes can be quite long and complex, allowing communication among multiple neurons at the same time. Neurons communicate with one another electrochemically: neurons activate, or "fire," via an electrical charge, but at the synapse, where the axon of one

neuron meets the dendrite of another, a chemical called a neurotransmitter is used to communicate across the gap. The axons
and dendrites are sheathed in a white, fatty substance called myelin, which provides insulation that helps to preserve the electrical impulse as it moves along the axon. Diseases such as multiple
sclerosis result when the myelin sheath is disrupted or damaged in
parts of the nervous system.

The cerebrum of the brain is divided into two large hemispheres, which communicate with each other through a thick band
of white matter called the corpus callosum (and some smaller pathways). Each hemisphere is divided into major structural regions
called lobes, which can be mapped more or less accurately by refer
ence to major sulci. The principal lobes are the frontal lobe, parietal lobe, temporal lobe, and occipital lobe. Each is associated with
particular functions, and each is divided into many smaller sub-
sections or sectors, but it is important to keep in mind that parts
of one lobe can be in intensive communication with parts of other
lobes, leading to the formation of functional brain networks.

The cerebral cortex is conventionally divided into two functional types. The primary cortex is directly involved with motor
control or input from the senses. The primary motor regions are
located in the frontal lobe, mostly along the central sulcus, one
of the major landmarks on the surface of the brain. (As most of us
are aware, one side of the brain controls the motor functions of
the opposite side of the body.) The primary sensory regions are
distributed throughout the cerebrum. The primary cortex associated with the senses of touch and position is located in the parietal lobe, while vision is centered in the occipital lobe, hearing in
the temporal lobe, and olfaction (smell) in the frontal lobe.

Most of the cerebral cortex is not primary cortex, but rather
consists of what is known as association cortex. Association cortex

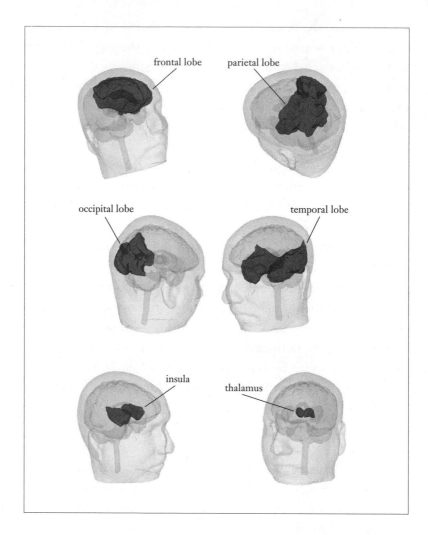

In each hemisphere, the human brain is divided into four major lobes: frontal, parietal, temporal, and occipital. The insula of each hemisphere is a buried "island" of cortex below the surface of the overlying frontal, temporal, and parietal lobes. It is particularly important for processing taste information, among other things. The thalamus consists of several nuclei (aggregations of neurons); in sum, these nuclei serve as the gateway to the cortex for information coming in from the body.

is aptly named in that it is where inputs from different parts of the brain associate with one another and primary information is processed. Some association cortex receives inputs from only one primary source, while others integrate different kinds of sensory information. It is generally thought that across mammal species, as brain size increases, the proportion of association cortex increases relative to primary cortex. Higher-level cognitive functions, such as thought, decision making, and creative endeavors, all originate in association cortex. However, it is important to keep in mind that these mental processes may occur with strong inputs from so-called lower-level brain regions. The limbic system incorporates several smaller structures distributed mostly along the inner surfaces of each of the hemispheres, buried deep within the midbrain of the cerebrum. The limbic system regulates many basic functions, such as the sense of smell, basic emotions, and memory. The expression "going limbic" to describe someone filled with anger or fury or lust reflects the notion that there is a primitive, animalistic part of the brain that occasionally emerges under emotional stimulation. Another group of buried midbrain structures are the nuclei of the basal ganglia. These structures are especially important in the control of movement in general, and in motivated behaviors such as those involving feeding.

As Antonio Damasio has forcefully argued, emotion and feelings are essential to classical higher-level processes such as decision making and consciousness.[21] The brain is an integrated structure, not a collection of more or less evolved regions nested one within another like a series of Russian dolls. This vertical integration is important to keep in mind when we consider how we think food and eating. You can pop an unleavened cracker into your mouth, chew, swallow, and digest it without thought or feeling. But if you are a Catholic in a church or cathedral during

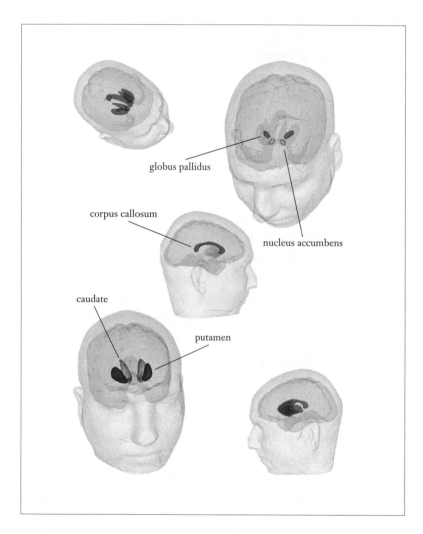

The basal ganglia are nuclei that include the globus pallidus, nucleus accumbens, caudate, and putamen. They are important for movement and in motivated activities such as feeding and ingestion. The corpus callosum is a white-matter structure located between the two hemispheres. The vast majority of communication between the two hemispheres is routed through the corpus callosum.

Mass, the unleavened cracker is a communion wafer, and in the process of consumption the wafer transubstantiates into the body of Christ, then there is very likely to be thought and feeling behind your actions. Transubstantiation is not amenable to scientific study, but recent advances in the neurosciences make it possible to understand the brain processes that occur in an individual as he or she undertakes devotional and other complex, culturally infused actions.

So, with the orientation tour of the brain out of the way, let's get back to how the brain regulates chewing, which is where eating ostensibly starts.[22] The muscles of mastication connect to the brainstem via the trigeminal nerve (cranial nerve V). The rhythmic pattern of chewing is controlled by a group of neurons in the brainstem known collectively as the central pattern generator (CPG). When you are chewing, the CPG receives inputs from higher-level areas of the brain while maintaining a complex feedback circuit involving the trigeminal nerve and other nuclei in the brainstem. Although there is higher-level control of chewing, experiments on animals that have had their cerebrums removed show that basic patterns of chewing can be maintained by the CPG and the trigeminal circuit alone. Despite this involvement in a very fundamental and evolutionarily primitive behavior, the CPG may also be part of circuits important in the production of speech, which is clearly a higher-level cognitive process. Again, there is an integration of the higher and lower parts of the brain.

So what is going on in these higher-level parts of the cerebrum when we chew? Researchers have used some of the most sophisticated brain imaging technologies, such as functional magnetic resonance imaging (fMRI, which measures changes in blood flow to different regions of the brain while the subject is doing different tasks), to try to pinpoint the parts of the brain that are active during chewing.[23] One problem has been that in order for these techniques to work their best, the head needs to be still, but chewing makes that somewhat difficult. Of course, older methods of brain research, such as lesion analysis (correlating the loss of a brain function to a specific brain injury site) or direct electrical stimulation during neurosurgery, had already localized the mouth- and tongue-related areas of the primary motor cortex of the frontal lobe and the sensory areas of the parietal lobe. These findings were confirmed in the earliest fMRI studies, which also showed expected activation in the cerebellum (voluntary motor control) and the thalamus (a collection of nuclei in the midbrain that acts as a critical relay station between the cerebral cortex and lower brain regions). In addition, there was activation in the insula, a small bit of cortex buried beneath the surface of the frontal and parietal lobes; the insula is a zone where integration of inputs from several cortical regions occurs, and it is implicated in the regulation of taste, among many other functions.

One fMRI study suggests that there may be a larger network associated with chewing that extends to parts of the association cortices of the frontal and parietal lobe. This was discovered by comparing the brain activity of subjects chewing gum versus those making jaw movements without gum (sham chewing).[24] What exactly about gum chewing causes activation in these areas is unknown. We might expect that other foods, with other associations, consumed in different contexts, would bring about activation in various association areas above and beyond those involved directly in the motor control and sensory monitoring of chewing.

Crisp Foods Are Noisy

Chewing crispy or crunchy foods should activate another functional network in the brain: the hearing network.[25] We hear by using specialized cells within our inner ear that can detect the movement of air and convert that to neural signals. We also hear by the conduction of vibrations through the bones of the skull, which can be detected by a structure within the ear. All this vibration detection from the ear is conveyed to the brain via cranial nerve VIII, which also carries fibers responsible for the ear's other functions, maintaining balance and detecting changes in the position of the head. The auditory fibers of cranial nerve VIII make their way to the brain stem, then up through various nuclei of the midbrain, and finally to the primary auditory cortex. This is located along the top surface of the temporal lobe—the "thumb" of the brain as you look at it from a side view. The primary auditory cortex is localized in and around a ridge in the cortex known as Heschl's gyrus. This is surrounded by association cortex for auditory processing. It should come as no surprise that some areas vital to the comprehension of spoken language are located in the temporal lobe association cortex near Heschl's gyrus.

When you chew something crispy (or anything, for that matter), taste and smell will be the dominant senses brought into play, and I will discuss these later. But sound is always there as well. For the most part, the sounds of eating that tend to concern people are the ones that everyone can hear. Table manners may focus on the elimination of the "gross noises" (as emilypost.com calls them) of slurping and smacking, or they may encourage the production of these sounds, as in the enthusiastic slurping done by Japanese noodle eaters. Restaurant owners and managers know that manipulating the aural dining environment, through music or modifying the acoustic qualities of the dining room, can influence how much and how long diners eat.

The most constant eating sounds we hear are, of course, those we hear in our own heads. But actually, these are typically the sounds we *stop* hearing. A common feature of all neural sensory systems is habituation—the responsiveness of sensory neurons typically decreases with continued exposure (which can vary according to circumstances) to a stimulus. When you put clothes on, you initially feel the fabric against your skin, but in no time you habituate to this very familiar sensory stimulus. Functional MRI shows that the activation of Heschl's gyrus and the surrounding association auditory cortex decreases with a repeated auditory stimulus, consistent with habituation.[26] This is interesting because the auditory signal passes through several neural pathways and relays before reaching the cortex, but habituation is still reflected at this highest level of auditory processing.

Habituation is necessary for the brain to perceive the surrounding environment in the midst of multiple sensory inputs; generally, more intense or unusual stimuli take longer to habituate to. Celebrated chefs such as Thomas Keller and Ferran Adrià (see Chapter 7) actively combat sensory habituation by serving many small, varied courses over the course of a long (and expensive) meal. This is nothing new: variety has long been a characteristic of celebratory feasts. This variety is typically seen as a display of wealth or plenty, as in the bounty of the American Thanksgiving meal. One reason it is so easy to overeat on Thanksgiving is that the number of courses served exceeds that of a typical American meal, and it is usually much easier to eat a smaller amount of many different foods than a whole lot of one food. The modern, industrialized food environment is historically unique, however, in that it provides an abundance of the kinds of highly appealing foods and flavors that resist habituation. For example, most people are willing to eat a large quantity of popcorn if it is loaded with butter and salt rather than left plain and dry.

Perhaps one reason that crispy foods have such an appeal lies in their ability to stimulate our hearing as well as our senses of taste and smell. Crispy in and of itself stands apart from other food qualities based on texture; this texture can be pleasurable even when combined with flavors that are themselves not necessarily all that appealing. Chewing crispy foods is louder than chewing non-crispy foods. If habituation takes longer given a stronger sensory signal, then we should enjoy eating crispy foods for a longer period of time during any given bout of eating. Of course, there are always numerous factors important in determining what we like to eat, but all things being equal (that impossible-to-achieve, thought-experiment state), it is not unreasonable to suggest that we might like a particular crispy food in part because we like the way it sounds in our own head. So the next time you eat a potato chip, when you savor the flavor, savor the sound.

The Crispy Word

Onomatopoetic is one of those big words that we learn as kids in school and tend to actually remember when we get older. It is a fun word to say, once you get the hang of it, and providing examples of it *(buzz, hiss)* pretty much explains what it means. We learn about all kinds of rhetorical devices in English classes, but while our knowledge of the difference between metaphor and simile might become hazy over time, onomatopoeia is much easier to retain. This likely has nothing to do with the grammatical and rhetorical importance of the concept expressed by the term: onomatopoetic terms are not all that commonly used, after all.

But by definition, onomatopoetic words are self-referentially evocative. This evocativeness serves to reinforce our knowledge of the concept.

Words such as *crispy* and *crunchy* are onomatopoetic. The etymology of *crispy* is somewhat complex (the first definition in most dictionaries is "curly or wavy"), but it is clear that whatever its ultimate origins, the word has come to be used primarily as an adjective for brittle foods. Now, it is clear that the sound of the word *crispy* is not identical at all to the sound that crispy food makes in our mouths, but for some reason it evokes that sound to our ears. Similarly, the word *crunchy*, which is widely acknowledged to originate in onomatopoeia, evokes an even more profound sense of this quality.

Crispy may be a better menu word because it tends to be used to indicate a more refined and controlled sense of brittle food, while *crunchy* is evocative of something louder but less processed and more wild. But why should either word enhance the appeal of a particular food or help to sell it in a restaurant? Onomatopoeia may be one reason. Research in functional brain imaging shows two distinct ways in which the words *crispy* and *crunchy* might evoke eating even before the food is consumed.

How does the brain react when it hears an onomatopoetic word? Naoyuki Osaka and his colleagues have conducted a series of fMRI experiments exploring this issue.[27] Osaka argues that this is a particularly important phenomenon in the Japanese language, as Japanese is rich in onomatopoetic words. Osaka and his colleagues have discovered that when subjects hear some onomatopoetic terms, they show activation in parts of their brain that may be active when they are actually experiencing the action or emotional state the term evokes. So, for example, the anterior part of the cingulate cortex (located on the inner surface of the hemispheres along the midline of the brain) is important in mediating the link between the centers of emotion in the limbic system and the executive functions of the frontal lobe. It is also one of the regions that is active when an individual feels pain. Osaka and colleagues found that when a subject simply hears a word that is suggestive of pain, there is activation in the anterior cingulate cortex. Other researchers have found that merely seeing a person with a painful facial expression can also activate this region.[28] Beyond anything having to do with onomatopoeia, it is quite interesting from a social cognition and empathy perspective to see that perceiving pain in others generates activity in the same brain networks that light up when we are experiencing pain ourselves.

Osaka also looked at brain activation in subjects when they hear a word that suggests walking. Six such terms were used in his experimental protocol, each of which denotes for Japanese-speakers a particular kind of walking: *teku-teku* (two different versions), *suta-suta*, *toko-toko*, *yochi-yochi*, and *yota-yota*. As a control, these were compared to a series of repetitive terms that were similar in overall phonetic feel but did not have any meaning. Osaka found that there was activation in part of the visual association cortex located near the primary visual areas in the occipital lobe (at the back of the brain). This is somewhat surprising since the subjects had their eyes closed during the experiment and therefore were not receiving any visual inputs at all; they only heard the terms. However, the part of the visual association cortex that was activated is one that has been implicated in the processing of visual information about body actions. Thus, simply hearing a word that suggests walking provokes the same response as a visual image of someone walking.

So onomatopoetic terms appear to have the power to promote brain activation in regions that monitor and experience emotions and those that are involved in mental imagery. But what is the relationship between the brain regions that monitor actions and those that are active when a person actually engages in the actions?

Since it is difficult to do very much moving within the confines of an MRI machine, researchers interested in brain activation of motor control, an important area with implications for rehabilitation and all kinds of physical performance training, have had to investigate the relationship between motor imagery (imagining doing an action) and the brain control of motor actions.[29] What they have found, for a wide range of motor actions, is that the primary motor areas of the brain are activated appropriately during motor imagery. In other words, thinking about doing an action activates the same parts of the brains that come into play when you are actually doing the action. Motor imagery is not completely the same as motor actions, since it is not clear that the primary sensory regions are activated during imagery. Nonetheless, the fact that there is a substantial functional congruence between motor imagery and motor execution allows fMRI and other neuroimaging techniques to serve as important tools for research in this area.

The implications for crispy and crunchy in all this now begin to take shape. Simply reading, hearing, or saying the onomatopoetic terms *crispy* and *crunchy* is likely to evoke the sense of eating that type of food. Presumably this feeling would be represented in the brain by activation of the mouth and tongue regions of the primary motor cortex (and of course, when a word is actually said, the motor regions of the mouth are being directly activated). *Crispy* might be such a compelling descriptive term because, in a sense, hearing or saying the word strongly promotes the motor imagery of eating—a food item with the word *crispy* attached to it is in some ways already being eaten by its potential consumer.

Crispy in a menu could be quite persuasive, especially when coupled with the fact that crispy foods are often quite palatable for other reasons. *Slurp* is another onomatopoetic eating term. In some contexts, such as in describing an iced beverage sold at a convenience store, it too could be attractive in the same sort of way that crispy is. However, we do not want to see *slurpable* on the menus of fancy restaurants. As we discussed earlier, slurping is an unacceptable sound at the formal Western dining table, and I suspect that even ordering a food with the word *slurp* attached to it would be difficult for anyone with any sense of table manners, no matter how enticing it otherwise appeared to be.

A Crisp Conclusion

Are we any closer to understanding the "something" that makes crispy food innately appealing? Why are we crazy for crispy? The human species has numerous ancestors and relatives for whom a crispy insect was and is an attractive meal. Even today, people in many cultures are quite happy to make a meal of a cricket, grub, or grasshopper. Some of our kindred species feast on raw, crispy vegetables, and even with those species for which leaves and stalks are not a first choice (and we humans would be in that category) a preference for them is quite useful if the need arises to survive on fallback foods. We have an evolutionary legacy as primates that suggests that crispy and crunchy foods should be attractive to us, at least sometimes and under certain conditions.

With the advent of cooking, dietary conditions changed drastically. Crispy became available to our ancestors via the Maillard reaction. Cooking made the nutrients in meat and certain plant foods, such as tubers, more available to us, and more palatable as well. Our ancestors who liked crispy cooked foods may have done particularly well in the reproductive sweepstakes, since cooking allowed greater access to a whole range of high-quality food items in varied environments. Our innate liking for crispy, derived from our distant relatives, may have been reinforced in more recent evolutionary times by the advantages conferred by cooking.

Crispy foods may, in various small ways, have a privileged place in the brain. Crispy foods incorporate hearing into the sensory mix of eating, and it is very likely that the stronger and more varied sensory mix provided by crispiness staves off boredom and habituation while we eat these foods. And as we just discussed, the word *crispy* itself may increase the appeal of such foods, at least when we are contemplating making them part of a meal. This would be an unexpected consequence of having a brain that is wired for language while still profoundly influenced by cognitive processes that occur well below this higher cognitive level.

There are other possible reasons why crispy may be so appealing, of course. In the modern food environment, commercially produced crispy foods are ubiquitous and strongly promoted— and, at the same time, demonized as leading to obesity. These foods, or at least some of them, are "bad." But as many of us are aware, some more so than others, doing something bad, as long as it is not too bad, can be pleasurable in and of itself. Eating a bag of potato chips may be enjoyable not just because it delivers ample salt, fat, and carbohydrate in a nice crispy package, but also because of the frisson of illicit pleasure it confers in a hectoring, contradictory nutritional culture.

Crispy foods are certainly not the only type of food that humans find appealing, and of course some people do not like crispy foods. But I begin with the generally appealing quality of crispy because in this book I set up a framework for exploring the human way of eating, with the ultimate goal of understanding some of the reasons people eat the way they do. How we think food and how we eat food are complex products of multiple histories. These cognitive, evolutionary, and cultural histories interact in unique ways in each individual, who brings to the table a personal history as well. These all combine to produce within each individual's brain a multifaceted "theory of food," which I will discuss in greater detail in the final chapter.

2

THE TWO-LEGGED, LARGE-BRAINED, SMALL-FACED, SUPEROMNIVOROUS APE

It is a remarkable circumstance, that man alone is provided
with a case of instruments adapted to the mastication of all
substances,—teeth to cut, and pierce, and champ, and grind; a
gastric solvent too, capable of contending with any thing and
every thing, raw substances and cooked, ripe and rotten,—
nothing comes amiss to him.

—Quoted from "Gastronomy" in *Hogg's Weekly Instructor* (1849)

ANIMALS SUCH AS GOATS AND PIGS might have reputations as
natural garbage disposals, but as this anonymous nineteenth-
century commentator pointed out, we humans do a pretty good
job of eating an extremely wide array of foods. As a species, we
have a basic biological proclivity to forage widely, to avoid specializing on a certain category of food. On top of this biology,
however, culture makes us selective, dictating which foods we
use and how to prepare them. The combination of cultural innovation and a biological predisposition toward generalized diets
results in an almost infinite universe of foods associated with the
human species. Of course, not all people eat all kinds of foods,
and in fact, many people do not eat a very wide variety of foods

at all. Part of our adaptability as a species is that individual members of our species can exist quite happily on very different diets. How did this come to be?

At least some of the answers to this question can be found in our evolutionary history.[1] An unusual ape appeared somewhere in Africa about 6 million years ago. This ape was different from other apes in that it spent less of its time in the forest and more in an open grassland or forest margin environment. It was also different because it walked on two legs instead of knuckle- or palm-walking on all fours like present-day great apes—orangutans, chimpanzees, and gorillas. Why this ape started to walk on two legs and move out of the trees is still not well understood. Many reasons have been suggested, ranging from energy efficiency to the ability to carry to the development of family units requiring provisioning, but none is generally accepted as *the* explanation.[2] Whatever the reason, this bipedal ape was successful, and its kind diversified into several species, spreading out over the course of millions of years to occupy much of the African continent, and subsequently Asia and the rest of the world. All of these bipedal ape species belong to a single zoological tribe, the Hominini, or "hominins" for short. Despite the fact that several different species and lineages of hominins evolved over the past several million years, *Homo sapiens* is the one and only hominin species that survives today.

Diet played a key role in the evolution of some of the diverse forms of hominins. One hominin lineage, which includes at least three species from East and South Africa, comprises the robust australopithecines. They earned the moniker "robust" not because their bodies were exceptionally large and heavy-set (in fact, they were about chimpanzee-sized) but for their relatively massive heads. Over the course of a couple of million years of evolution, the robust australopithecine skull was completely reworked from a more generalized ape version to one that accommodated large muscles for chewing; the molar teeth enlarged greatly, and the front teeth (incisors and canines) shrank to almost trivial size. The robust australopithecines had a nickname, "Nutcracker Man," that seems to have been well earned. Those massive jaws and teeth look perfect for crushing hard food items such as nuts and seeds.

But anatomical first impressions can sometimes be misleading. Recent research on robust australopithecines has focused on what their teeth can tell us about how they lived.[3] Clues about their daily diets can be gained by microscopic analysis of wear patterns on teeth; in addition, the enamel of the teeth contains carbon isotopes that yield insights into the plant foods they were consuming. What these studies show us is that the big teeth and jaws of the robust australopithecines evolved not for chewing hard objects but rather for eating grasses and sedges (a group of grass-like plants). The South African robust australopithecines ate a more ape-like diet of fruits and vegetation combined with the grasses and sedges, while the East African group looked to be almost wholly dependent on grasses and sedges. So it was a diet of grasses, not nuts and seeds, that drove natural selection to produce such an extreme and apparently specialized skull.

The first robust australopithecines appear in the fossil record around 2.5–2.7 million years ago. The period around 2.5 million years ago was a very interesting time in our evolutionary history: not only do the robust australopithecines appear in eastern and southern Africa, but the oldest stone tools date to this time and place as well. Then, around 2 million years ago, hominins with enlarged brains rather than enlarged teeth also start to be represented in the African fossil record. Climate fluctuations at this time led to changes in the relative amounts of forest and grassland present in the continent, which in turn had a clear effect on the type and distribution of animals.

Among our human ancestors and cousins, a clear split developed in the family tree. On one branch we have the large-toothed, robust australopithecines, whose species are now classified into the genus *Paranthropus* and which have no living descendants. On the other, we have the hominins of the genus *Homo*, whose brain sizes clearly exceed those of the great apes and earlier hominins; this branch eventually leads to the evolution of hominins with brains about triple the size of those seen in chimpanzees (that would be us—and some extinct forms such as Neandertals).[4]

The last of the robust australopithecines, the highly specialized East African forms, went extinct about a million years ago; they ultimately appear to have become overspecialized for grass feeding. Being a specialist can be great, but if conditions change or competitors arise, shifting out of the specialized mode can be difficult. By contrast, their cousin species *Homo erectus*, our ancestors, became omnivorous, using their increased intelligence and enhanced technology to take advantage of a variety of habitats (grasslands, lakesides, mixed woodlands) and expand the range of foods they could eat.

But how were our ancestors able to do this and avoid the trap of overspecialization? What were the critical factors in our evolution that allowed our ancestors to not only make their way in their African homeland but ultimately expand outward into extraordinarily diverse environments throughout the world? One critical factor was intelligence, sheer brainpower. But since brains are part of bodies, an omnivorous brain evolved along with an omnivorous body. The modern human body is basically a primate body with some modifications. So let's look back briefly at the origins of primates to see where our dietary evolution started.

Eating in Trees

As a rule, primates spend all or most of their lives in trees. There are exceptions (most notably ourselves, some baboon species, and large-bodied great apes, who spend a considerable amount of time on the ground), but they are few and far enough between that we can be comfortable with this generalization. Going back 60 million years or so, all apes, monkeys, and a motley collection of small primates known as prosimians (lemurs, tarsiers, lorises, and bush babies) shared a common ancestor that probably looked more like a small rodent than like any primate alive today.[5] It was at this time, at the end of the Mesozoic Era and the beginning of the Cenozoic, when we see the changeover from the age of reptiles to that of mammals. Mammals existed in the Mesozoic, but they were not major players on the scene (although the split between placental and marsupial mammals dates to this era). A major catastrophe, perhaps an asteroid or comet striking the earth, is thought to have led to a period of substantial global cooling, killing much of the plant life on earth, which in turn spelled the end for the large plant-eating dinosaurs and the carnivores that preyed on them.

The demise of the reptiles provided an opportunity for mammals. At the same time, there was a diversification of flowering plants (angiosperms), which depended largely on insects for pollination. This interplay between plants and insects set the stage for the development of modern ecosystems and provided the foundation for the diversification of placental mammals into the orders (primates, rodents, carnivores, etc.) we see today.

Nearly a century ago, anthropologists looked at the unique anatomical features that set primates apart from other mammals and noted that many of those features seem like adaptations for life in the trees. Since most primates live in trees, perhaps this was not the most insightful observation. But researchers looked at the grasping hands and feet with nails instead of claws, and the reliance on vision over smell (especially stereoscopic vision, with depth perception, which is provided by forward-facing eyes), and concluded that the primate body evolved primarily for negotiating tree branches in a complex three-dimensional environment.[6] These anatomical adaptations were what allowed the earliest primates to separate themselves, both literally and in an evolutionary sense, from the other mammals. What made primates special was not so much what they were doing but where they were doing it. This became known as the arboreal theory of primate origins.

The arboreal theory held sway for decades, and anthropologists and primatologists were very comfortable with the idea that all primate characteristics could be explained by tree living. Beginning in the 1970s, however, researchers started to take another look at primate origins. Matt Cartmill pointed out that there are lots of other mammals that live in trees but do not possess grasping hands or feet or stereoscopic vision (think squirrels); thus it was unlikely that these primate traits evolved exclusively for tree living.[7] Instead, he suggested that they fit the profile of a visual predator, one that anchored itself with its grasping feet while it snatched insects with its free hands, using the depth perception afforded by stereoscopic vision to better home in on its prey. Robert Sussman linked the movement of primates into the trees with the expansion of angiosperms.[8] He argued that they were following fruit, not just insects, and that grasping hands and feet would have been useful for reaching fruit on small branches, as well as grabbing insects. So the traits of primates had to do not just with where they were living but also with what they were eating in those trees.

The fossil record of 45–65 million years ago is replete with a variety of more or less primate-like critters, some of whom were undoubtedly the ancestors of today's living primates.[9] Most of the earliest ones are quite small, which is important because a mammal that eats only insects can't weigh much more than about 500 grams, a little over a pound, unless it can tap into entire colonies of insects, as anteaters and aardvarks do. So it is quite possible that the earliest primates could have survived on an insect-only diet. It is clear, however, that over time, primates, especially anthropoids (monkeys and apes), became increasingly dependent on plant food, including fruits and gums, and for some species large quantities of leaves (the ability to digest leaves evidently evolved independently in both Old World and New World monkeys).

So if we were to say that primates live in trees and subsist mostly on plant food, we would not be too far from the truth. There is one wholly carnivorous primate, the tarsier, a small prosimian that actually belongs on the monkey and ape side of the primate family tree rather than with the lemurs, lorises, and other prosimians. But among monkeys and apes, only the smallest monkeys get even as much as a third of their daily food intake from animal sources. The vast majority get almost all of their food in the form of plants. Among our closest relatives, gorillas and orangutans are almost 100 percent dependent on plant foods, while for chimpanzees the figure is slightly smaller but is typically over 90 percent.[10] Humans stand apart in this regard: diets in traditional hunter-gatherer populations typically are reported to be about 65 percent plant foods and 35 percent animal foods, though there can be substantial variation in this figure depending on local conditions.

We humans don't live in trees, and on average we eat a lot of meat, at least compared to our monkey and ape cousins (there of course is much individual and cultural variation in the amount of meat humans eat). We know that the advent of bipedality signaled a move away from the trees and forests and into more open grassland. The robust australopithecines made this move and remained plant eaters. Our kind took a different route out of the trees, eventually incorporating increased amounts of meat into the diet while still eating a variety of plant foods. We also evolved a very large brain, which quite likely helped us take over the open grassland niche from the robust australopithecines. But what did eating meat have to do with this? To rephrase a classic evolutionary conundrum, which came first, the meat or the brains?

Brains and Meat

The evolutionary model of "man the hunter/woman the gatherer" has been around in various guises since the first half of the twentieth century, and debates about what it means for gender roles and relations date to the 1970s. At its core, the model recognizes the importance of meat in the diet of our ancestors and the critical place of hunting in their lifeways. As a pioneering advocate of the hunting model, Sherwood Washburn, wrote in 1957:

> The taste for meat is one of the main characteristics distinguishing man from the apes, and this habit changes the whole way of life. Hunting involves cooperation within the group, division of labor, sharing food by adult males, wider interests, a great expansion of territory, and the use of tools.[11]

One point that Washburn was trying to make was that meat was critical for *human* evolution—evolution of members of the genus *Homo*. In other words, meat was not important for the separation of the hominins from the apes, but came into play only later, with increasing brain size, stone tool use, and enhanced intelligence and cognition. Washburn was directly contradicting the views of the South African paleoanthropologist Raymond Dart, discoverer of the first australopithecine in 1925 and inventor of the genus name *Australopithecus*. Dart hypothesized that the trend toward meat-eating had started with the small-brained australopithecines rather than with the larger-brained members of the genus *Homo*.[12] To Dart, the australopithecines were "killer apes" who used a variety of non-stone implements to kill animal prey and even other members of their own species. Washburn, and later C. K. Brain, who made a definitive study of the bone deposits in South African hominin fossil sites, argued strongly that the australopithecines were most likely the hunted rather than the hunters, and that the reason the bones of South African australopithecines were found among the remains of prey animals is that they too were deposited in caves or sinkholes by large cats, hyenas, or predatory birds.[13] This conclusion is now widely accepted, and thus increased meat-eating is seen as a hallmark of the evolution of later hominins only (dating from 2 million years ago).

Washburn tied the eating of meat directly to the development of hunting, specifically the cooperative hunting of large game by males. During the 1960s, this developed into the "hunting hypothesis" of human evolution—the notion that hunting was the paramount step toward early humans becoming more like modern humans. The hunting hypothesis spurred intensive research on the few remaining hunter-gatherer groups still living a traditional life in the late twentieth century. Although no such group lives in stasis, frozen in a cultural evolutionary past, these groups certainly maintained lifestyles that were closer to the conditions under which humans evolved than was the case for anyone living in a developed, agricultural society. The hunting hypothesis came under some withering critiques in the 1970s and 1980s, both by feminists, for being male-centric, and by some archaeologists, who argued that many of the bony remains of animals discovered in hominin archaeological sites could be interpreted as being the result of scavenging by hominins rather than hunting.

Let's consider two points on the timeline of human evolution. At the first point, around 2.5 million years ago, we have hominins who were largely dependent on plant materials for their diet; at the second point, 15,000 years ago, we have modern human hunter-gatherers who ate a diet that was still mainly plant-based but also includes a significant amount of meat. If we go back in time from 15,000 years ago, there is ample evidence that the large-brained Neanderthals (who lived from about 30,000 years ago to 150,000 years ago) were quite effective large-game hunters; preceding them, the mix of hominin species in the period 300,000 years ago to 1 million years ago also shows archaeological evidence of hunting.[14] Indeed, archaeological evidence indicates that as long ago as 1.75 million years, with the appearance of early (but not the earliest) members of the genus *Homo* in East Africa, hominins were using stone tools to butcher and process large animal carcasses.[15] More critically, archaeologists have found that they ate the choicest parts of the animals, which means that they had access to the carcasses while they were fully intact. This pattern could be a result of "power scavenging"—say, if hominins chased carnivores off the carcass shortly after the carnivores did the killing—or (and this might be the simplest interpretation) of hominins being the actual hunters. In either case, there is strong evidence for the increasing importance of meat in the diet of *Homo* reaching back to almost 2 million years ago. The evidence for increased meat in the diet is almost coincident with the beginning of the trend toward increased brain size in *Homo*. So some time in the period 2 million to 2.5 million years ago, the earliest members of genus *Homo* likely began the transition from being almost fully plant-eaters to being much more omnivorous.

There is no way of knowing if early *Homo* hunted in the same way as modern humans do; it seems highly unlikely. But the basic technology of defleshing bones with a sharp stone leaves the same signs no matter how the animal was killed. Washburn's scenario of how humanness derived from hunting is still difficult to place in the timeline, as many of the higher-order behaviors associated with hunting are difficult to identify in the archaeological record. Nonetheless, as Henry Bunn and Craig Stanford write, "current evidence indicates that the acquisition and consumption of meat may not have made us hominins, but there is compelling evidence that meat-eating had a major, influential role in making us human."[16]

The thing that ultimately separates us from other primates is, of course, our behavior—which reflects our intelligence, language, and other cognitive abilities. This is not simply the result of increased brain size, since there are also changes in the functional organization of the brain. However, brain size increase is what we see in the fossil record, and there is no doubt that increasing brain size is associated with enhanced intellectual processing power. So if hunting, or meat eating, played an influential role in making us human, then it very likely had an influential role in the evolution of a larger brain. Why should this be the case?

The Expensive Brain

Brains are hungry. The reason for this is that brains are largely composed of neurons, and these neurons communicate with each other in a way that requires a significant amount of cellular energy. The signal that passes down the axon of a neuron to communicate with another neuron is called an "action potential." This is an accurate name, since the action potential requires the movement of ions across the boundaries of the cell at the synapse, the space between one neuron and another. In fact, neural tissue uses sixteen times more energy than skeletal muscle. In order to meet the neurons' energy demands, mitochondria, known informally as the "powerhouses of the cell," are found in high concentration at synapses.[17]

The increase in brain size over the course of human evolution has been far from trivial.[18] Nearly all of the great apes and the non-*Homo* hominin species have brain volumes in the range 350–550 cc. The earliest members of *Homo*, those living around 2 million years ago, typically had brain volumes in the 600–850 cc range. Not long after this, *Homo erectus* emerged, first in Africa and then in Asia. Brain size for this species ranged around 900–1,200 cc, with a general upward trend over the course of its more than 1 million years of history. Starting several hundred thousand years, a group of specimens appeared in Africa, Europe, and Asia that are sometimes referred to as archaic *Homo sapiens* (or more formally as *Homo heidelbergensis*), which had brain sizes in the range of 1,100–1,400 cc, bigger than *H. erectus*. Finally, around 200,000 years ago, Neandertals and modern humans appear on the scene, with skulls capable of holding 1,400 cc of brain, and sometimes considerably more.

Rates of brain size increase in the human lineage significantly outpaced increases in body size; thus our brains were becoming not only absolutely but also relatively larger. Compared to other mammals, the human brain puts an unparalleled strain on the body's energy resources. The human brain accounts for only

2 percent of the body's mass, but a whopping 20–25 percent of the body's resting metabolic rate. The figure for other primates is in the 8–13 percent range, while for non-primate mammals, 3–5 percent is typical.[19] How did our ancestors support this increasingly large brain? Well, as you might expect, it comes down to meat, though the story is rather complicated.

For all animals, dietary options are limited to plants and other animals, and a species's nutritional well-being depends on achieving a salubrious (for that species) ratio of plants and animals in the diet. Most primates do very well eating mostly or only plants, and despite the fact that the human species as a whole can be characterized as omnivorous, individual people can certainly thrive on an all-plant diet. While animal foods are nutrient-dense, high in protein and many minerals and vitamins, this is true of many plant foods as well. What may have been important in our particular evolutionary history was the relative availability of plants and animals in a specific environment. One critical factor is that there appears to be no barrier to adding meat to a generally non-meat-eating primate's diet: experimental studies on primate digestion have shown that most primates can easily handle, and even relish, animal food as well as plant food.[20]

Most primates live in the trees. In the forest canopy, however, meat comes in small packages and can be difficult to catch. Although primates may depend on some animal foods for some nutrients, they are generally better off devoting their time to plant foods that are more accessible and take less energy to obtain. On the grasslands of the savanna, the situation is different. Animal food is readily available and comes in potentially very large packages, if a predator can figure out how to get it. Primatologist Katharine Milton suggests that by incorporating meat into their diet, along with energy-rich plant foods such as fruits, nuts, and starchy roots, early *Homo* could have avoided eating low-quality plant foods (leaves, stalks), which must be consumed in large volume, require a large amount of gut space for digestion, and take time to obtain in adequate quantities.[21] Meat contributed to the high-quality diet necessary to support a large brain, in part by allowing our ancestors to focus on higher-quality plant foods at the same time.

Brains and guts seem to be inextricably, evolutionarily linked. In 1995, anthropologists Leslie Aiello and Peter Wheeler wrote a classic paper that put the focus on gut size in the debate about the evolution of diet and brain size.[22] Aiello and Wheeler paid attention not simply to diet but also to the anatomical and physiological trade-offs within the body that could have made it possible to physiologically support a larger brain. These trade-offs were necessary because the increase in brain size occurred without an increase in the body's basal metabolic rate. Aiello and Wheeler began by noting that the brain is not the only expensive tissue in the body: the heart, kidney, and splanchnic organs (liver and gastrointestinal tract) use up more than their fair share of energy as well. By analyzing organ and body sizes in a wide range of primates, Aiello and Wheeler sought to determine how big these various organs would be expected to be in a primate our size. They found that our hearts, kidneys, and livers were about the expected size; however, the gastrointestinal tract was 60 percent smaller than expected.

The amount of energy saved by this reduction in the size of the gastrointestinal tract was just about the amount needed to offset the increased energy demand imposed by a larger brain, suggesting a straight gut-for-brains energy trade-off. Aiello and Wheeler argued that given the oxygen demands of a larger brain, there could not practically be a reduction in the size of the heart or lungs. Similarly, since the brain cannot store energy in the form of glucose, it is dependent on a steady supply of it produced by the liver, so a large brain probably also could not tolerate much in the way of liver reduction. And the hot savanna climate in which early *Homo* evolved put a premium on kidney function to maintain urine concentration; thus kidneys were not a candidate for a reduction in size. An evolutionary scenario thus emerges: the higher the quality of the diet, the shorter the digestive tract can be, and the greater the proportion of the body's metabolic activity that is available to maintain a larger brain. There is a circularity here, of course, but that does not mean that the interplay of these factors was not critical for human evolution. Aiello and Wheeler's analysis has not gone without criticism in the scientific community, but it certainly has spurred research in this area, and the idea of anatomical trade-offs for increased brain size is generally accepted.[23]

Primates with a higher-quality diet tend to have larger brains than those with a low-quality diet. Meat is not a necessary component of a high-quality primate diet.[24] However, it appears very likely that our ancestors improved the quality of *their* diets by increasing the quantity of meat they consumed. They may have increased the quality of their diets in other ways as well. As discussed in the previous chapter, cooking probably increased the efficiency with which meat, along with calorie-rich but hard-to-consume plant foods such as tubers, could be processed and digested.

But hunting animals clearly has an important place in the sociocultural fabric of traditional hunter-gatherer groups (in many instances this cultural importance even seems disproportionately large in comparison with the amount of meat in the diet). While the deep antiquity of the cultural importance of meat is difficult to assess, it seems reasonable to assume that the cultural status of meat and hunting was built on a nutritional foundation. Reduced gut size and increased brain size are anatomical markers of nutritional and cultural (and cognitive) revolutions. Meat very likely had a critical role in mediating the causal relationships among these revolutions.

The Fish-Brain Hypothesis

The advantages of meat in the diet for human evolution are often expressed in general terms. Meat provides a readily available (to a savanna-dwelling hunter), high-density source of calories, protein, and other vitamins and minerals, and the adoption of a meatier diet is thought to have allowed a shrinking of the gut to provide that critical energy trade-off we discussed in the previous section. But no specific nutritional component of meat is thought to have been essential for the evolution of increased brain size or higher cognitive functions. Fred Previc has suggested that in going to a meatier diet, early *Homo* increased the amount of protein they were consuming on a regular basis.[25] This potentially gave them access to greater quantities of the amino acid tyrosine, which is a precursor to the neurotransmitter dopamine. Previc argues that many dopamine pathways in the brain important for higher-level cognitive and language functions became elaborated during human evolution, and a meatier, higher-protein diet would help maintain these pathways. Although the tyrosine-dopamine connection might be important for brain function, any increases in tyrosine in the diet probably only conferred a secondary advantage, however.

Large grassland-dwelling, meat-providing grazers were not the only sources of animal food that became available when our ancestors moved out of the trees. Lakes and streams provide a ready and nutritious source of fish and other aquatic critters that are totally unavailable in the forest canopy (not many streams and rivers up there). Is it possible that incorporating fish into the diet may have provided a jump start toward the evolution of a larger brain in our ancestors? An evolutionary conundrum identified by Stephen Cunnane and Michael Crawford is that if early *Homo* depended on technology to obtain meat, which in turn required a somewhat larger brain and more sophisticated cognitive abilities, how did these early humans initially support a larger brain if they were not cognitively ready to obtain a high-quality diet?[26] What change in dietary behavior could support the evolution of a larger brain without initially requiring a great increase in intelligence? Perhaps it was fishing.

Unlike the argument for the importance of meat, the fish-for-brains hypothesis relies not on general nutritional quality but on access to certain dietary substances found in aquatic foods.[27] Specifically, advocates of an aquatic-diet theory focus on the fatty acids that are essential components of developing nervous systems. They suggest that relatively small-brained hominins ate fatty-acid-rich aquatic foods, which supported the development of a larger brain (which they could then use for sophisticated hunting, cooperative behavior, and so on). The most essential of these fatty acids are thought to be docosahexaenoic acid (DHA) and arachidonic acid (AA). Although AA is available from egg yolks, organ meat, and muscle meat from land animals, the best sources for DHA are fish and shellfish (AA is also present in aquatic animals). Crawford, Cunnane, and their colleagues hypothesize that early *Homo* species exploited the shallows of African lakes and rivers, where a potential abundance of fish and shellfish could be obtained. They argue that this would not require a technological advance but rather should be seen as an expansion of traditional gathering techniques. Thus aquatic foods could have provided a jump start for cognitive evolution without requiring a cognitive revolution.

This is an interesting idea, placing the origins of hunting not out on the savanna, with undersized but clever hominins doing battle with large prey and vicious predators, but on the weedy shorelines of rivers and ponds. The aquatic food hypothesis does have some issues, however. The most notable one is that essential fatty acids are available from other food sources, or can be synthesized in the body from readily available substances.[28] Another compelling argument against it has been that there is very little archaeological evidence for the exploitation of aquatic food by early humans. While one could argue that ancient low-density populations might not leave much of a mark on the landscape and that the absence of evidence is not the evidence of absence, even so—and in contrast to the vast shell middens that have been left by modern human populations along various seashores—the African archaeological record of 1–2 million years ago is not supportive of extensive aquatic food consumption.

Outside Africa, however, the archaeological evidence may be changing. Recent studies suggest that early hominins may have indeed exploited marine resources. Coastal-dwelling Neandertals living 40,000 years ago in Gibraltar (near Spain) almost certainly ate marine animals. Within a cave site, archaeologists led by Chris Stringer identified an ash layer containing a hearth, Mousterian stone tools (almost always associated with Neandertals) and knapping flakes, and an abundance of mussel shells derived from a nearby estuary.[29] This short occupation site provides us with a nice snapshot of Neandertal life. As Stringer and colleagues write: "This occupation level . . . records several activities in the life of the Neandertal occupants. These activities consisted of selection and collection of mollusks, transport of the gathered mussels to the cave shelter, fire making in the cave, the use of heat to open the shells, consumption of these mollusks, knapping on the hearth embers, and subsequent abandonment of the site."[30] Deeper excavations in the same area yielded the remains of seals and dolphins from even earlier occupations, along with the bones of terrestrial mammals more typically associated with Neandertal hunting. A few fish remains were also found in these deposits.

The Neandertal evidence from Gibraltar demonstrates once and for all that modern humans are not the only hominin species who ate from the sea (or lake or river). But this finding does not say too much about the antiquity of seafood use—the Gibraltar Neandertals were hanging on at the periphery of the original Neandertal range at a time when modern humans were the dominant hominin species. Stringer and colleagues even speculate that these Gibraltar Neandertals may have been able to survive longer than their inland kin because of their access to both land and water resources.

So if 40,000 years is just a drop in the bucket in terms of the broader picture of human evolution, can we find any seafood in the archaeological record that gets us closer to the more critical 2-million-year window of the aquatic hypothesis? Evidence from a river site in Java may do just that. In the early 1890s, the Dutch army surgeon Eugène Dubois discovered the first remains of the species we now call *Homo erectus* in Java, at a site called Trinil, located on the Solo River. Dating of this site is somewhat controversial, but the time frame for it is in the range of 900,000 to 1.5 million years ago; as it is today, Trinil was then a riverine environment, not all that distant from lakes, deltas, and the sea. Dubois and subsequent investigators collected a large quantity of material from these sites, including the skeletons of lots of fish, mollusks, mammals, birds, and reptiles, plus the remains of a few human ancestors.

This collected material was reanalyzed by José Joordens and her colleagues.[31] Carefully picking through this collection, they found at least eleven edible mollusk species and four fish species that would be obtainable from shallow water; a hominin with little technology could have done quite well grazing in these waters. But did *Homo erectus* take advantage of this aquatic larder? To answer this question, Joordens and colleagues looked at the distribution and size of the shell remains. Two particularly abundant mollusks presented a somewhat intriguing picture. First, rather than being distributed evenly throughout the site, their remains were concentrated in one layer and area. Second, almost all of the specimens were larger adults, with juveniles noticeably absent. This was not because the original fossil excavators ignored small or fragmentary material: they were quite meticulous in this regard. Rather, Joordens and colleagues hypothesize that *Homo erectus* may have been the ones choosing the larger adult mollusks, eating them, discarding their shells in a limited location, and creating a shell midden to be discovered a million years later.

The discovery of aquatic food exploitation by Neandertals and evidence of the possible systematic consumption of shellfish by *Homo erectus* certainly expand the horizons in time and space of seafood eating in hominins. The evidence doesn't take us all the way back to the origins of the genus *Homo* in Africa, but it does demonstrate the ways the hominin diet may have expanded long before modern humans appeared on the scene. Critics of the aquatic food hypothesis are correct in that there are other sources of fatty-acid-rich foods around; by the same token, meat from terrestrial sources does not provide anything that cannot be provided by plant food. However, advocates of the hunting hypothesis have provided not just a theoretical justification for the increase of meat in the diet but also evidence of this increase in a particular evolutionary setting. So even if meat is not a physiological necessity, it was likely a critical player in the evolution of the human brain and cognition. More evidence is needed before we can say the same about aquatic foods. On the other hand, it is looking increasingly likely that increased brain size and intelligence in *Homo* went along with developing an expanded and more varied diet, which included animal foods not only from the land but from water as well.

From Omnivory to Superomnivory

Maybe it's time to retreat from meat, land-based or otherwise. After all, the increase in animal consumption led to a higher-quality diet that in most cases still relied on plants for the majority of its calories and nutrients. The evidence of meat eating signals the development of a technologically based omnivory, rather than one based on expanding food resources through specialized diet-related anatomical features. The development of sophisticated technology has been just one facet of human cognitive evolution, however. Behavioral plasticity and flexibility are part and parcel of increasing intelligence. The ability of our ancestors to explore the environment, productively test novel foods, and communicate information about food to members of their social group were just as important as any advances in tool making or use. Aquatic foods were probably not necessary to make a bigger brain, but our evolution was undoubtedly abetted by our ancestors' willingness to try new foods, including those found near and under the water.

About 1.8 million years ago, members of the genus *Homo* left Africa for the first time. They migrated northward through what is now the Middle East and proceeded to fill much of the Old World, reaching Europe, China, and Indonesia. Dietary versatility must have been a critical factor in this diaspora of *Homo erectus*.[32] They must have been able to maintain a high-quality diet in the face of novel environments, seasonality, and natural cycles of feast and famine. Cooking probably helped them to exploit new foods, and stone tools, as well as tools made of wood or hide that are not preserved in the archaeological record, provided them with certain advantages over animals that may have been better adapted to specific environments.

These ancestors of ours were also probably helped by the shared traditions and knowledge they possessed in their social groups—in other words, their culture. Now, anything we have to say about how *Homo erectus* lived or went about their business on a day-to-day basis can only be speculative. But we know that some chimpanzee groups possess practices and even tools that are not seen in other groups.[33] These traditions can be sustained over time by observational learning. This type of behavior is often referred to as protocultural rather than cultural, but that may suggest a difference more of degree than of kind, although some might want to reserve the term *protocultural* for traditional transmission in the absence of language.

At any rate, *Homo erectus* were undoubtedly cultural animals. We do not know if they possessed language or something like it, but they must have had a communication system that contained some elements of our own form of spoken language. Their increased brain size suggests an enhanced ability to store information, which would have been available for sharing with others in a social group. Given the diversity of environments in which

H. erectus groups lived, there was undoubtedly variability in diet among these populations. But beyond ecological variation, diet probably also varied along cultural lines, as different successful practices were perpetuated over time in different lineages.

Omnivory in *H. erectus* may have signaled the beginning of the kind of omnivory we see in our own species—that is, an omnivory quite different from the sort represented in the ecologist's data table of x percent plants, y percent meat, and z percent seafood. Cultural anthropologist Jon Holtzman has conducted extensive research on the foodways of the Samburu, a traditionally pastoralist group living in northern Kenya. His description of the Samburu diet in its cultural context illustrates the fully human approach to how and what we eat:

> Samburu cuisine presents a seeming conundrum. On the one hand, what and how one eats is central to the complex construction of the most integral relationships and values in Samburu life. Food and eating practices are crucial to social action and the symbolic world, and the types of food one eats, the context for eating, and the company with whom one eats construct crucial aspects of individual and group identity across the lines of ethnicity, kinship, gender, and age. On the other hand, Samburu diet is sparse, ideally constructed of just three livestock products: milk, the daily staple; meat, a desirable supplement available mainly a ritual occasions or when an animal happens to die; and blood, a kind of quasi-milk, consumed as a boost to vitality or a supplement during times of scarcity. Granted, the real Samburu menu is rather more complex.[34]

Furthermore, what the Samburu do *not* eat is almost as culturally significant as what they do eat:

Dietary restrictions also serve to construct ethnic boundaries. Samburu prohibit the consumption of a wide range of
potentially edible terms—for instance, fish, reptiles, birds,
donkeys, and many game animals. By far the strongest prohibition is against eating elephants, which are considered
to be similar to human beings. Such prohibitions help construct ethnic boundaries with Turkana pastoralists and
Dorobo foragers.[35]

Like all human cultures, the Samburu have their own unique
cultural history and diet. Nonetheless, they can serve as an exemplar of the human way of omnivory. In all human societies, food
and eating are embedded in a cultural web involving status, kin,
and identity. Even during times of food stress and famine, human
cultures do not descend into anarchy.[36] When food shortages occur, cultures exhibit a fairly stereotypical set of responses. There
is initially an intensification of shared activity, as alternative
foods are sought and stored. This is followed by a social retreat, as
people individually begin to hoard food and withdraw from common activities; there is an increase in violence and aggression over
food sources. Inevitably, some cultural practices and institutions
break down as the famine continues. Nonetheless, Peter Farb and
George Armelagos point out that the "skeletal structure" of cultural systems is preserved even in the wake of serious famine,
which they argue is "evidence of the intimate linkage of food with
culture and society."[37]

Human omnivory goes beyond the basic notion that we are
a species that gets food from a range of sources. As we've seen,
what we eat depends on our evolutionary history—a specific physiology adapted to a certain set of environmental conditions. An
additional factor is each individual's personal history of experiences and preferences. However, there is another history that

comes into play: that of the culture in which an individual is born, raised, and lived.

Alfred Kroeber, one of the pioneer American anthropologists, conceived of culture as "superorganic."[38] He thought that the way to understand culture was to see it as both organic and something beyond organic. In the same sense that organic entities possess qualities that transcend those of inorganic substances, Kroeber argued that some of the hallmarks of culture, such as transmissibility, high variability, value standards, and so on, cannot be explained in terms of the organic composition or behavior of individuals. Culture is superorganic and superindividual in that it is transmitted by learning without becoming part of the genetic endowment of individuals. Culture persists and changes above the level of the individual or even groups of individuals within a culture. Nonetheless, the "organic endowment" of human beings, along with the various laws of physics and nature, places real constraints on cultural phenomena.

Kroeber's concept of culture as superorganic was first proposed in 1917 but never really found a following in anthropology, for a variety of reasons.[39] Still, I think that when it comes to human eating behavior, the concept works rather well. Although we can talk about cultural foodways and what different foods mean in different contexts, ultimately there are always biological factors underlying diet that cannot be ignored by even the most insistent cultural determinist. However, to simply say that humans are omnivores is really inadequate for conveying the nature, complexity, and diversity of human diets. I would argue instead that humans are superomnivores. Human diets are the product of our biological history, but their diversity both within and across groups is largely (not entirely) a function of variability at the superorganic level. Understanding how and what people eat as a species cannot really be done without acknowledging that diets appear and disappear, diversify and converge, expand and contract, all as a function of cultural factors.

In this regard, the diversity of human diets starts to sound a lot like the diversity of human languages. Languages are cultural entities that exist and evolve at the superorganic level. But language itself is a biological feature of the human species, and its expression depends on the proper functioning of the brain and various other parts of the body. We can still study the neurocognitive basis of language even if languages are cultural phenomena. Similarly, human eating behavior has a biological basis that exists as the foundation for our species-wide superomnivory.

Agriculture: Superomnivores on Restricted and Unrestricted Diets

If there is anything that makes humans superomnivores, totally removed from more conventional zoological notions of diet, it is agriculture. Before the advent of agriculture, which arose in various locations throughout the Old and New Worlds beginning about 10,000 years ago, human hunter-gatherers procured and processed food using technologies that were largely elaborations of techniques that we could see employed in somewhat similar forms in other animals. Even cooking, which may have been a transcendent technology in the context of hominin evolution, is basically an externalized form of chemical processing, which all animals employ to digest their food.

With agriculture, humans enter into a synergistic or coevolutionary relationship with another species. Such species-species interactions are common in the zoological world: the bacteria we carry in our guts, for example, help us digest certain carbohydrates for which we do not possess the appropriate enzymes. However, agriculture is different in that humans alone initiate the relationships with other species, and human needs dictate the course of animal and plant domestication. Now, in an evolutionary sense, this may be to the great benefit of the species we domesticate. Corn and other grains have spread throughout the world from their local origins, "exploiting" humans as a vehicle to evolutionary success. But humans are the conscious, selective agent of their success. This makes agricultural synergy or symbiosis quite different from any other form.

Agriculture led to increasingly large population centers, civilization, empires, and ultimately the technologically advanced world of the twenty-first century.[40] According to many commentators, it was the worst thing that ever happened to our species. The "noble savage" has long been celebrated in arts and letters, but for centuries a simplistic notion of cultural evolutionism, in which civilization and civilized people clearly outranked those of a more primitive bearing, held sway. In the first part of the twentieth century, anthropologists worked hard to tear down this essentially prejudiced and wrongheaded view of cultural variation.[41] But it perhaps wasn't until the 1960s that the old ladder of progress was turned on its head. It was around this time that Marshall Sahlins famously described hunter-gatherer groups as "the original affluent society."[42] Based on empirical data from some of the few remaining hunter-gatherer groups, Sahlins argued that they worked less and had more leisure time than their contemporaries in developed economies. Although they had few possessions, they were not poor; as Sahlins stated, "Poverty is a social status. As such it is the invention of civilization."

In the dietary realm, Sahlins noted that large numbers of people in agriculture-based societies went to bed hungry each night, while hunter-gatherers had a steadier and more varied food supply. This food supply was based on intensive knowledge of the seasonal availability of plant and animal foods in the environment. Historically, such knowledge was not obvious to all "civilized" observers of aboriginal hunter-gathers, although some did note that hunter-gatherers managed surprisingly well despite being handicapped by a decided lack of civilized knowledge.

The varied diet of hunter-gatherers stands in stark contrast to the monotonous diet of many agricultural populations. In some cases, reliance on a single food crop can lead to diseases resulting from vitamin deficiency.[43] Populations in which corn is a staple can be susceptible to high rates of pellagra, a disease caused by a deficiency in the B vitamin niacin. The symptoms of pellagra can be quite unpleasant, including a distinctive rash, diarrhea, and even mental disturbances. Some cultures traditionally treated corn with an alkali, which released niacin from the hull of the corn, thus reducing the likelihood of developing pellagra. An overdependence on polished rice can lead to beriberi, a disease of the nervous system caused by a lack of vitamin B_1 (thiamine).

More generally, traditional agricultural diets, because they are less varied, are not as good as hunter-gatherer diets at providing all of the specific nutrients that our bodies need to thrive. On the other hand, they clearly provide enough to allow people to survive and reproduce, increasing population numbers. Many archaeological studies have been done comparing the skeletal health of groups before and after the advent of agriculture in the same location. In almost all cases, the agricultural group shows signs of nutritionally based bone or tooth stress not present in the hunter-gatherer comparison population.[44] In addition, agricultural groups, because of their higher population densities, are more likely to be exposed to infectious diseases. If domestic animals are present, then there is also a risk of disease being transmitted from animals to humans.

Things have changed considerably in developed countries with the shift from traditional to modern industrialized agriculture. Food is relatively cheap and abundant, and the supply is buffered from the effects of seasonality and poor yields. Initially, combined with medical advances to control infectious disease, modern agriculture supported the growth of healthier and larger populations; people were also individually healthier and of bigger physical size. In the mid-twentieth century, however, it became apparent that there were groups that did not do so well with this modern Western diet. For example, when some Native American and Pacific Islander populations gave up their traditional agriculture-based diets (which also included fishing and hunting) for a Western diet, they became highly susceptible to developing obesity and all of the medical conditions, especially diabetes, associated with it. Geneticist James Neel came up with the idea that non-European populations were at risk for these diseases precisely because they had not been previously exposed to such a flush nutritional environment.[45] He argued that in European populations, natural selection had eliminated the "thrifty genotypes" that were in effect more metabolically efficient but which increased the risk of developing obesity and diabetes in a rich nutritional environment.

Neel was probably wrong in thinking that premodern European diets were particularly abundant, and so his scenario for the historical elimination of thrifty genes in those populations is probably wrong also.[46] Nonetheless, the high susceptibility of some recently Westernized groups to developing obesity and diabetes was and is very real, for whatever reason. And it is beginning to look as if these populations were the canary in the coal mine. Today, across the developed world, and in developing nations with emerging middle classes, we see rates of obesity and diabetes increasing almost yearly. Modern lifeways combined with modern diets are apparently making people fatter and less healthy. Ultimately, agriculture can be blamed for this state of affairs, so it is not surprising that some people argue that we can solve these dietary problems by going backward, to practices that predate the advent of agriculture.

Modern Paleolithic Superomnivores

There are critics of the dismal view of agricultural life and the conversely rosy view of traditional hunter-gatherers.[47] Yet there is also clearly a strong sense that as much was lost as was gained with the advent of agriculture, at least in terms of diet. The relative affluence of developed countries gives their residents options. One option is to dietarily re-create the past, specifically the Paleolithic (which encompasses human history before agriculture, all the way back to the appearance of stone tools 2.5 million years ago). Ironically, the modern market economy and industrialized, scientific agriculture give people the opportunity to "eat like a caveman" if they want to.

As evidenced by this list of diet books, the Paleolithic option is being taken up, or at least considered, by quite a few people:

> *The Paleo Diet: Lose Weight and Get Healthy by Eating the
>   Foods You Were Designed to Eat*
> *The Paleo Diet for Athletes: A Nutritional Formula for Peak
>   Athletic Performance*
> *The Paleo Solution: The Original Human Diet*
> *The Primal Blueprint: Reprogram Your Genes for Effortless
>   Weight Loss, Vibrant Health, and Boundless Energy*

The Primal Blueprint Cookbook: Primal, Low Carb, Paleo,
  Grain-Free, Dairy-Free, and Gluten-Free
The Evolution Diet: What and How We Were Designed to Eat
The Evolution Diet: All-Natural and Allergy Free
The New Evolution Diet: What Our Paleolithic Ancestors Can
  Teach Us about Weight Loss, Fitness, and Aging
The Paleolithic Prescription: A Program of Diet and Exercise
  and a Design for Living
NeanderThin: Eat Like a Caveman to Achieve a Lean, Strong,
  Healthy Body
Orthomolecular Diet: The Paleolithic Paradigm
Health Secrets of the Stone Age

Paleolithic diets seem to have spawned their own little publishing industry. These books first started to appear in the late 1980s, not long after S. Boyd Eaton and Melvin Konner published the first formulation of a Paleolithic diet in 1985.[48]

So what exactly did Paleolithic peoples eat and how does it differ from what most people eat today? It is important to keep two things in mind. First, when we talk about a Paleolithic diet versus a contemporary diet, we are referring to the idealized average diet of an idealized average individual. Of course, there was and is much variation due to all sorts of factors. Second, Paleolithic peoples were simply much more active than people are today. They needed more energy to get through their daily activities.

Compared to a contemporary diet (i.e., the one found in Westernized, developed countries), the Paleolithic diet is higher in calories but gets fewer calories from fat and more from protein.[49] In fact, the levels of protein consumption are much higher than those recommended today. The Paleolithic diet is much higher in fiber and micronutrients. There is much higher consumption of potassium relative to sodium. Calories from carbohydrates make up similar percentages in both diets. However, Paleolithic eaters took in carbs in the form of fruits and vegetables, which are also rich in fiber and micronutrients, while people in developed countries load up on carbs from refined cereal grains and simple sugars, which are not rich in either.

Although the modern diet is vilified, it is important to remember that most of its negative effects are felt later in life. Heart disease, diabetes, and cancer are the diseases of populations that are fortunate enough to have limited exposure to infectious disease, adequate medication to fight the infections that do occur, and more than enough calories to make sure that babies grow and thrive. Even in populations with high rates of obesity, diet does not prevent people from reaching adulthood and reproducing, even if their overall health is not up to ideal standards. Nonetheless, there is evidence to support the "mismatch hypothesis" of Paleolithic bodies living in an environment for which they are not optimally suited. As Staffan Lindeberg, a clinician who has extensively studied the effects of Paleolithic eating in the modern urbanized world, writes: "Humans are apparently not well adapted for staple foods that were introduced less than 10,000 years ago, including cereal grains, milk, salt, and refined fat and sugar. . . . Human physiology seems well designed for a mixture of lean meat, fish, shellfish, insects, and a large variety of plant foods."[50] In clinical trials, Lindeberg and his colleagues have shown that a Paleolithic-type diet can reduce symptoms of type 2 diabetes more effectively than either more conventional diets aimed at diabetes control or the famous Mediterranean diet.[51]

Leaving aside the question of whether the Paleolithic diet is the best choice for people today, it is important to recognize that the modern agriculturalist diet is good at appearing diverse (look at the breakfast cereal aisle in the supermarket) while at the same time recycling a limited array of basic foodstuffs. The luxury that people in developed countries have is that a very diverse diet truly can be obtained at the supermarket, if there is a willingness to stray from the most common products of industrialized agriculture. Although what we eat is still a reflection of social status and is molded by cultural institutions, individuals in developed countries have more opportunity than ever to pursue a diet of their choosing. This may mean a vegan or gluten-free diet, or a Coke and doughnuts for breakfast every day. We may also choose to supersize our superomnivory.

Advocates of the Paleolithic diet like to say that it reflects 99 percent of the time we have lived as members of the genus *Homo*. Of course, it only covers about 3 percent of our time as primates. It was over the course of primate evolution that our buttons for sweet, salty, fatty, and even crispy, for wanting more food rather than less, evolved. We all still respond when those buttons are pushed. Even if it is clear that our bodies do better with a Paleolithic diet, our minds do not necessarily want to go along. There are several reasons for this. One is that, as a species, humans have been dietary opportunists, flexible feeders able to adapt to a wide range of environments. The Paleolithic diet was to some extent a contingent diet; when the constraints are removed, we do not necessarily choose to re-create it. Agriculture has its own advantages, including the ability to support larger populations and the powerful cultural institutions they create. The social forces unleashed by agriculture have shaped not only our diets but our minds and bodies as well.[52] Finally, at the individual level, the Paleolithic diet is a diet like any other—a conscious change from the habits that developed over the course of a lifetime. As we will discuss later, changing a diet can be among the more difficult things a person can do.

Superomnivory means that all manner of diets are "natural" for us. Of course, we use culture to achieve superomnivory; from a strictly nutritional perspective, that means many of the decisions we make are "unnatural." Perfectly edible foods are rejected because they are considered "unclean," or because they are eaten by the enemy, or for no crystallized reason at all. There surely must have been an evolutionary cost to be paid for such capriciousness. At the same time, those costs must have been outweighed by the benefits of embedding food and eating in wider cultural and cognitive systems. We see hints of this in other primates: feeding patterns may be strongly influenced by dominance hierarchies in monkeys, and chimpanzees exchange food (meat) for sexual access. But humans do more than any other primate to make food something that is less about nutrition and body maintenance and more about social living. This is the basis of our superomnivory and ultimately a key to the evolutionary success of our species.

3

FOOD AND THE SENSUOUS BRAIN

In Paris the gourmets eat with quiet deliberation, rolling each mouthful slowly toward their gullets. In Jacques' little cottage three or four friends inhale the stew's rich fumes, and eat it down like the hungry workingmen they are. In Paris and in the village, there is a gusto, a frank sensuous realization of food, that is pitifully unsuspected in, say, the college boarding-house or corner café of an American town.

In America, we eat, collectively, with a glum urge for food to fill us. We are ignorant of flavour. We are a nation taste-blind.

—M. F. K. FISHER, *The Art of Eating* (Wiley Publishing, 2004)

PERHAPS AS A NATION Americans are a bit less taste-blind today than they were some seventy-five years ago. The romance of the TV dinner has passed, fast food is widely acknowledged to be rather less than a mixed blessing, and consumers increasingly look for some hint of quality or nature even in the most highly processed packaged foods. America is now a land of celebrity chefs, food-oriented game shows, an expanding network of farmers' markets, and more cookbooks than one could cook from in a life-time.[1] Yet at lunchtime the drive-through lanes of the local McDonald's or Wendy's overflow with cars holding solitary drivers who have the "glum urge" to fill their bellies in the least amount of time possible.

Food-focused Americans can take some solace in the fact that French eating patterns have also changed over the past decades. The French are no strangers to fast foods or convenient, pre-cooked meals. The preeminent status of traditional French cuisine has been challenged both from within the nation and from without by advocates of a lighter, more ingredient-focused approach incorporating influences from diverse culinary cultures. So perhaps the French diet and the American way of eating are converging a bit, with each food culture taking on some of the characteristics of the other. And yet, during a recent stay in Normandy, I more than once saw young men in their twenties dining alone, savoring multicourse lunches in a bistro or cafe with a glass or two of wine. Such a sight would be extremely rare in America, but in general the French invest more in their food than do Americans—more time, more money, more of themselves.

A vital part of investing oneself in food is an embrace of the sensory experience of eating. Taste is critical to this. We have two concepts of taste—a narrower notion referring to the physiological ability to perceive certain substances in our environment with our taste buds, and a broader concept that refers to the sum total of an individual's preferences and desires for specific foods. This broader sense (so to speak) of taste incorporates not just physiological taste and smell but also the other senses, such as touch, vision, and even hearing. Beyond that, taste in the broad sense develops out of individual experiences, reflecting both familial and cultural environments. Underlying it, however, is basic sensory biology.

Evolutionary studies of diverse animals and organisms show that much of the sensory machinery is shared, at least at the level of the cell and molecule. From the housefly to the house cat, the basic building blocks of the senses have been modified and repurposed to fit the different sensory needs of different species. For example, the visual systems of insects and vertebrates share commonalities such as the use of a small number of neuron cell types that are divided into subtypes, each of which consists of multiple cellular layers. Visual information is mapped in different species' nervous systems in similar ways through the various levels of visual processing. There are also similarities in how genes control the development of invertebrate and vertebrate visual systems. When it comes to the sense of smell, which is the ability to detect chemical substances (odorants) in the environment, both vertebrates and invertebrates rely on the interaction of these odorants with receptors in specialized olfactory cells; the odorant receptors are all derived from one large family of proteins. The sense of touch also seems to have a common molecular basis in worms and vertebrates such as ourselves.[2]

Taste Cultures

When we look across the wide world of multicellular organisms, we see at the most basic levels evidence for a shared origins of sensory systems. But if that's so, then why don't the French and the Americans—demonstrably members of the same species, after all—employ the sense of taste in the same way? Why is there gusto on one hand and glumness on the other? M. F. K. Fisher's observation cited at the beginning of this chapter has nothing to do with physiology. Taste is, well, a matter of taste—an exemplar of the subjective and unverifiably personal. Individual food preferences are shaped by the cultural environment, like all other aspects of human behavior. And attitudes toward food and eating are influenced by a host of factors that lie beyond the reach of taste receptors.

So why were Americans historically reluctant to fully embrace the sensory experience of eating? Americans have traditionally harbored a suspicion of culinary sophistication: plain food was better than fancy food, or at least it was healthier and more honest. This suspicion, David Kamp writes, could have arisen from any of several sources, including "this country's Puritan origins, its early inheritance of British culinary stodginess, or just a general don't-tread-on-me stubbornness."[3] In the nineteenth century, even as the influences of French and other more sensual culinary traditions were beginning to make their mark on American cuisine, there was a countering movement toward "scientific" forms of eating, led by the likes of the great masticator Horace Fletcher and the cereal king John Harvey Kellogg. Even the pioneering cookbook writer Fannie Farmer, whose very name has become a byword for the homely life, preached the benefits of laboratory knowledge in shaping the American diet.

These people's work signaled the beginnings of the American form of nutritionism (the term was coined by the sociologist Gyorgy Scrinis and popularized by food journalist Michael Pollan).[4] The ideology of nutritionism is founded on several principles or beliefs: (1) that food is best understood as the sum of its nutrient components; (2) that experts, such as nutritional scientists, are necessary to uncover the hidden realities (beneficial and malevolent) of food; and (3) that the main point of eating is to maintain the body and its health. In nutritionism, foods are seen simply as delivery systems for their component parts. In Pollan's view, this leads to nutritionism's most nefarious corollary—that there is no distinction between whole foods and processed foods. All that matters is the chemical composition of the substances we put in our mouths. The quantification of the components of food has its uses, of course. It would be hard for a city dweller on a Paleolithic diet to try to figure out the supermarket equivalent of a gazelle haunch, a handful of grubs, and some tubers all on his or her own.

If American tastes were destined to be shaped by cultural forces that ignored or actively disparaged a sensual engagement with food, then what made the French such culinary sensualists? Naturally, there were multiple factors. The years just prior to and just after the French Revolution were a critical period in the development of French gourmandism.[5] First, in the mid-eighteenth century, there was the evolution of the restaurant (the word itself was derived from the name for a restorative bouillon). In contrast to inns and taverns that offered a limited assortment of foods presented with indifferent service, often family style, the restaurant presented food in a refined environment and focused not just on the eater's hunger but also on his or her individual taste.

Historian Rebecca Spang identifies two other wider cultural trends that emerged from the ferment of the French Revolution, both of which helped shaped French sensory attitudes toward food.[6] At the turn of the nineteenth century, as the French Republic was transitioning into the First Empire of Napoleon, aesthetic debate on a variety of subjects was widespread and encouraged by government officials. There was a "bread and circuses" component to this official encouragement: according to the censors of the Ministry of Police, if the people were following debates about the arts and sciences, that meant that they were not looking too closely at the affairs of state or the economy. Food and eating were on the officially sanctioned list of "pleasures" that were eligible for public discourse. However, gastronomy was favored over agronomy as a topic for discussion: developments in the grain trade were censored, while reporting on the latest restaurants, festive table settings, and new pastries was encouraged. The aesthetics of food gained, or was granted, equal social billing with that of painting and music.[7]

The spirit behind one of the rallying buzzwords of the French Revolution, *égalité*, also found representation in the new French gastronomy. Alexandre Balthasar Laurent Grimod de la Reynière was the first restaurant critic, a prolific writer on gastronomy, and the most famous eater in all of early nineteenth-century France. In Grimod de la Reynière's view, the role of the gourmand was something new, transcending traditional social classes. The ability to appreciate food, both sophisticated and simple, was not something that automatically came to people just because they had a boatload of money. Spang writes, "In the meritocratic land of the gourmands, there were to be no venal office holders and no inherited titles."[8] The status of the gourmand was forged in the relationship between the eater's palate and his or her plate. Elevating the status of individual tasting ability became the foundation for a food culture in which the taste and presentation of a food are prized along with its ability to sate one's hunger. French cooking is, of course, not unique in this regard, but in France the historical record allows us to trace the origins of some of these attitudes.

At the beginning of their existence as modern nations, the United States had equality and France had *égalité*. Their food cultures reflect these common ideological underpinnings, but in diametrically opposed ways. In the United States, equality meant that food, which was relatively plentiful in comparison to the situation in much of the Old World, became a social leveler. Since it was not particularly valued, the sensual experience of food, apart from satiety, descended toward the lowest common denominator. In France, the ability to taste food, evaluate it, and communicate about it became a vehicle for social mobility. This in turn helped to raise the status of food within the culture, and fostered a more active and intense formal food culture.

Different cultural foodways have the power to shape how individuals within those cultures think about food, and ultimately their perception of it. Historian Massimo Montanari writes, "Taste is a cultural product. . . . The organ of taste is not the tongue, but the brain, a culturally (and therefore historically) determined organ through which are transmitted and learned the criteria for evaluations."[9] Taste is cultural, embedded in webs of traditions and practices. If we look across the human species as a whole, much of the variation we observe in humans' relationships with food can certainly be explained by differences in culture or ethnicity. But cultural variation is not the only kind of variation that we observe. Within cultures we find that some people taste better than others, so to speak. Of course, history and tradition are at work here, at the level of individual and family environment, but biology likely also plays a role.

Jean Anthelme Brillat-Savarin, Grimod de la Reynière's nearly exact contemporary and the most distinguished writer on food to emerge from that period, believed that while everyone could enjoy the art and pleasure of eating, not just anyone could achieve the exalted status of a gourmand: "I believe in inborn tendencies. . . . People destined to gourmandism are in general of medium height; they have round or square faces, bright eyes, small foreheads, short noses, full lips and rounded chins. The women so predisposed are plump, more likely to be pretty than beautiful, and have a tendency towards corpulence."[10] Such a perspective is consistent with the assertion that being highborn is not what makes one a good eater. The meritocracy of eating, of being able to employ one's senses in the service of identifying good food, is a function of the ideology of equality, while at the same time providing evidence for the assertion that equality does not mean that everyone is the same. Brillat-Savarin's linking of the physiology of taste with the morphology of the body reflects the science of his age, but the science of our age has provided evidence that the sense of taste varies constitutionally, just as vision or hearing varies from person to person and throughout the lifetime of an individual.

Taste Basics

We have seen how people's taste behavior is influenced at the culture-historical and even ideological levels. Let's look at the other side of the story—the molecular and physiological basis of taste, which we all more or less share. All of the senses depend on specialized cells that respond to stimuli in the environment by initiating a nerve impulse. Taste and smell both depend on detecting the presence of chemicals; hence the receptors for them are called chemoreceptors.[11]

Tasting something begins with the activity of specialized taste cells in the mouth (no surprise there). These taste cells are organized into small structures, consisting of 50–150 cells (including both taste cells and supporting cells), called taste buds. The taste buds are organized so that the chemoreceptors, which are located on finger-like extensions of the cell wall, face out into the oral cavity. As food is eaten, it is liquefied by saliva and chewing, causing the release of chemicals and improving access of those chemicals to the taste buds. (For that reason, as many people have noticed, a dry mouth impedes the sense of taste.) Taste buds are distributed across the surface of the tongue in bumpy structures known as papillae. These papillae contain about 10,000 taste buds in total. Taste buds are also found on other surfaces in the oral cavity. The taste cells in a taste bud are not sensory neurons (unlike olfactory cells); instead, they must synapse to a sensory neuron in order to cause a nerve impulse.

In contrast to smell, which depends on thousands of different receptors, taste is derived from a combination of five primary taste categories: sweet, sour, salty, bitter, and umami (or savory). Umami is the taste we associate with monosodium glutamate (MSG), a flavor enhancer often used in Asian cuisines; glutamate is an amino acid found in many protein-rich foods. I grew up on a diet of typical American foods liberally sprinkled with MSG. Looking back, I realize now that the flavors I associate with many savory foods, especially beef, were in part a result of the potentiating power of MSG. Some evidence suggests that "fatty" (the taste of fatty acids) may be a stand-alone taste as well, although more evidence is needed to confirm this.[12]

Out in nature, umami and sweetness both signal that the food being consumed is nutritious, while bitter taste is typically a signal of toxicity. Our first instinct upon tasting something bitter is to spit it out. Chemoreceptors for the different primary tastes are distributed at different concentrations on different parts of the tongue: sweet receptors congregate at the tip of the tongue; salt receptors are also found at the tip and at the upper front portion of the tongue; sour receptors line the back margins of the tongue; bitter receptors are concentrated at the back of the tongue. Umami receptors are congregated at the back of the mouth, at the entry to the pharynx. Information from the different taste receptors converges in the gustatory centers of the brain, where a unified taste perception forms based on the relative strength of inputs from the different receptors.[13]

The perception of taste does not simply rely on the taste receptors. The texture of a food can influence its taste; for example, a dry food may produce less flavor than a moist food. In addition, physiological factors can alter our desire for a particular kind or taste of food. One such specific hunger is "salt appetite." When sodium levels in the body are low, a craving for salt can develop, in effect making salty substances taste better than they usually would. As mentioned above, smell also plays an important role in taste perception. Chewing food releases chemicals that are picked up by olfactory receptors, and so it is the combination of the activation of both taste and smell receptors that produces the perception of taste.

Adaptation or desensitizing in taste is a complex phenomenon.[14] Short-term adaptation of taste receptors is like other forms of sensory habituation (such as the way we become habituated to the feeling of clothes on our skin), in that repeated exposure can lead to desensitization of the taste cells. However, since the exposure to a food stimulus as it passes through the mouth and down the gullet is usually of short duration, it is possible to eat many bites of a food without becoming habituated to its taste. But over the long term, adaptation to certain tastes forms as a consequence of repeated exposure and personal experience, memories, and conditioning, all of which can be strongly influenced by the cultural environment. When people start drinking coffee, they often recoil at the taste—the bitterness is signaling that this substance should not be allowed in the body. Over time, however, coffee drinkers learn to ignore this message and coffee can become an essential part of a daily morning ritual. The bitter-tasting chemical is still there, but coffee drinkers have adapted to it and are thus able to appreciate the other qualities of coffee's complex flavor and the other benefits that may be derived from consuming it.

The Tasting Brain

Tasting occurs in the mouth, but the perception of taste happens in the brain. The connections from the taste cells to the brain are relatively straightforward.[15] Neurons that connect to the taste cells in the mouth pass via cranial nerves (numbers VII, IX, and X, to be precise) to the brain stem. Here, taste contributes to the carrying out of some involuntary reflex actions, such as swallowing and coughing. We have all probably had the experience of encountering a bad or simply unexpected taste and spitting out whatever caused it, without thinking about where the spit may be flying. This demonstrates clearly that taste can be processed at this subcortical level, under the level of conscious awareness. Other taste fibers continue upward, however, through the relay station of the brain, the thalamus, and continue to the cerebrum. They eventually reach the gustatory cortex, which is located in the insula (the "island" of cortex buried beneath the frontal, temporal, and parietal lobes) and in part of the frontal lobe that lies over the insula. The amygdala, one of the subcortical structures critical for emotion processing, also receives sensory inputs important for evaluating taste.

The processing of taste information gets a bit more complicated at this point. Fibers from the gustatory cortex project to a region on the lower part of the lateral surface of the frontal lobe—this is called the orbital region of the frontal lobe, or the orbitofrontal cortex.[16] Here, specific neurons respond not only to taste stimuli, including neurons "tuned" to the five basic tastes, but also to specific odors that are associated (via learning) with tastes and to the "feel" of food in the mouth (this involves qualities such as viscosity, texture, hardness, and so on). Other orbitofrontal neurons specialize in combining information about taste, odor, feel, and even visual appearance. As Edmund Rolls, the foremost modeler of the brain's taste and olfactory processing networks, writes, "These combination-selective neurons provide an information-rich representation of a wide range of the sensory

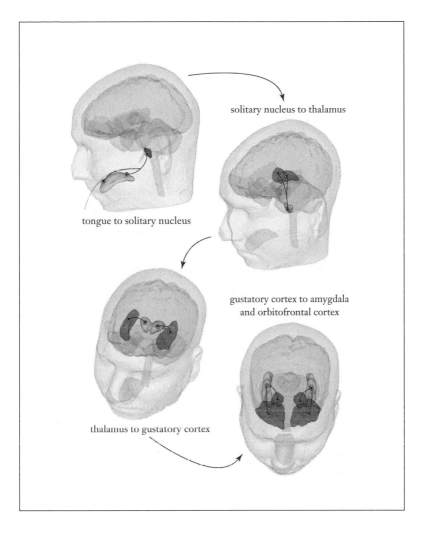

solitary nucleus to thalamus

tongue to solitary nucleus

gustatory cortex to amygdala
and orbitofrontal cortex

thalamus to gustatory cortex

The passage of taste information begins at the tongue, and then moves along cranial nerves VII, IX, and X to the solitary nucleus of the brain stem. From there it goes to the thalamus (gateway to the cortex) and then to the gustatory cortex of the insula and overlying frontal lobe. Taste information next passes to the amygdala and orbitofrontal cortex for more processing and integration with other kinds of information.

qualities of food."[17] Eating is truly a multisensory experience, and the dedicated neurons of the orbitofrontal cortex provide the functional foundation for this intimate sensory convergence.

One way that some diners have been able to more consciously experience the sensory convergence involved in eating is to remove one of the senses from the equation. An interesting phenomenon at the turn of the twenty-first century is that of the "dark" restaurant, in which, as one might expect, diners enjoy the experience of eating in complete darkness. This idea originated in Switzerland, where a blind clergyman, Jorge Spielmann, started a dark restaurant where the patrons could experience for a short time what it was like to be blind.[18] This concept, which can trend toward marketing gimmick, has been copied by restaurateurs in Europe, North America, and Asia. The emphasis is less on sharing the experience of blindness and more on exploring the supposed heightening of the senses of taste, smell, and touch that occurs when sighted people are deprived of vision. Dana Salisbury, owner of Dark Dining, a catering/performance company specializing in eating-in-the-dark experiences, puts it succinctly: "Why in the dark? Because it awakens the senses and presents new pleasures."[19]

Reading customer reviews of some of these dark restaurants on the Internet, one discovers that eating in them can be an unsettling experience, at least in mechanical terms (i.e., getting food from the plate to the mouth). But many patrons do say they experience a heightening of their nonvisual senses. The combination-selective neurons of the orbitofrontal cortex may provide an explanation for why this should occur: vision has a dedicated neural role in the perception of taste. Removing the visual input fundamentally changes how taste is perceived, increasing the relative contributions of other senses. In contrast, while ambient sound surely plays a role in the dining experience by influencing the diner's mood or ability to attend to eating, it should not have a primary role in influencing taste perception, because there do not appear to be combination neurons in the gustatory cortex (although there are sound inputs into the orbitofrontal cortex).

Research on the activity of single cells shows that the neurons of the orbitofrontal cortex are also critical in assessing the pleasantness of food.[20] Studies of the activation of single neurons in the taste pathway of monkeys have demonstrated that the identity and intensity of a taste are represented in the brain separately from its pleasantness. In other words, we don't have to be hungry to evaluate the taste of something, but when it comes to enjoying that taste, it helps to be hungry. This is why we often say that foods taste better to us when we are hungry and less good as we become sated. Neuroimaging research conducted on human subjects indicates that the orbitofrontal cortex is the brain's center for feeling pleasure while eating.[21] However, the intensity of a flavor is processed in other parts of the taste pathway.

Habituation decreases the pleasure that a person derives from tasting something, but what about the converse? One of the most interesting taste phenomena is that of potentiation or synergism: combinations of flavors make for a perceived taste that is more than the apparent sum of its parts. Salt and vanilla are well-known taste potentiators, but perhaps the most celebrated example of taste synergism in Western cuisine is the pairing of wine with food.

If we can chart anxiety about a subject by the number of instructional sites on the Internet, then it appears that there is quite a bit of confusion over just how one pairs wine with food, or at least anxiety about making the wrong choices. The basic principles of wine pairing go back to the early part of the nineteenth century, when they were codified in celebrated chef Alexandre Viard's *Le Cuisinier Impérial* (1817). This included classic pairings such as white wines with fish and oysters and vintage Burgundy with a roast. Viard and his contemporaries also advocated starting a meal with Madeira or sherry, a practice that later French chefs, such as Auguste Escoffier, dismissed because these strong wines overpowered the wine served later in the meal.[22]

So what goes on in the brain when someone perceives an enhanced flavor from consuming foods in combination? Although there do not seem to be any studies of subjects sipping wine and munching on food while in an MRI machine, there has been research based on combinations of potentially potentiating flavors. In a very interesting functional MRI study, Edmund Rolls and members of his research group tracked the brain activity of subjects who tasted MSG and inosine monophosphate (IMP), another umami flavor enhancer, separately and then in combination (a glucose solution was also tasted as a control).[23]

Rolls and his colleagues found that both MSG and IMP were considered pleasant-tasting on their own. Like glucose, both MSG and IMP caused activation along the whole taste pathway, including the orbitofrontal cortex. (The glucose solution, which was rated as the most pleasant of the three substances, also activated the anterior cingulate cortex, a region of the brain important in motivation and emotion-related higher cognitive processing.) The somewhat surprising result in this study came when the MSG and IMP were tasted in combination. It was already known that MSG and IMP in combination can result in taste synergism. The research participants confirmed that the combination of MSG and IMP was more pleasant than either separately; more significant, imaging showed that the combination activated a region of the orbitofrontal cortex to a much greater degree than would have been predicted from adding together the individual activations from MSG and IMP. In other words, the subjective taste synergism was accompanied by a spike in activation in the part of the brain that assesses the pleasantness of a taste.

This study cannot prove that pairing certain wines with certain foods potentiates taste in the same way as MSG and IMP work together. Some wines work with food simply because they do not clash with the food and are good enough to drink without being offensive. Basic table wines, lacking flavor complexity, can fit this bill. But wine pairing with food can on occasion jump to a different sensory level, when true taste synergy apparently occurs. For example, sweet wines can go quite well with sweet foods, but many people are surprised that some cabernet sauvignons work exceptionally well with dark chocolate: the chocolate flavor tones of the cabernet serve to potentiate the taste of dark chocolate. The neuroimaging research demonstrates that the aha moment of tasting a wine well paired with food is likely not simply a gourmand's affectation but an enjoyable neurological phenomenon.

It may be that taste synergy contributes to the superomnivory of the human species. Other animals may in the moment of consuming a combination of flavors derive increased pleasure from that combination. But only humans can talk about these combinations with others, and remember and elaborate upon them as part of family and cultural traditions. More foods become palatable if they are enhanced when consumed in combination with other foods. As we saw in the last chapter, agricultural diets based on one or two staple cereal crops are notoriously less varied than hunter-gatherer diets, but the blandness of the basic agricultural diet can be mitigated by the addition of small quantities of other foods and flavors to the staple cereal. This could not be considered an advantage of agricultural diets over hunter-gatherer diets, but taste synergy may have been one factor that helped enhance the quantitative advantages of agricultural foodstuffs by overcoming possible qualitative shortcomings.

The Pleasure of Pain: Eating Hot Peppers

Pain is not a taste; pain is pain. The nerve cells that act as pain receptors are the same in the mouth as elsewhere in the body. Some are fast-acting and thus more responsive to acute injury, warning the individual that he or she should take quick action to avoid more pain. Other pain receptors respond more slowly, reacting not to the immediate insult but to chemicals released when tissues are damaged. This sort of aching pain can persist long after the episode that caused the acute pain is over; aching pain can also result from tissue damage caused by disease or inflammation.[24]

Unlike other kinds of sensory receptors, pain receptors generally do not adapt to stimuli very quickly. As we discussed earlier, eating large quantities of the same thing, no matter how good it is, typically reduces the responsiveness of the taste cells. However, as many of us know from personal experience, eating more and more of a spicy hot food leads not to habituation but to more and more pain. Just as there is little habituation to eating something that is thermally hot, spicy hot foods do not cause the pain receptors to slow or shut down. The spicy hotness of a food is therefore a sensory quality quite distinct from its taste.

Over the long term, of course, people can habituate to hot food. I had a housemate back in undergraduate days who came from a cultural background that did not include very hot foods. Goaded on by other housemates, he once reluctantly put a tiny quantity of hot salsa on his morning eggs. He thought it was very hot, yet over the course of a few months he adjusted to the spiciness, dumping progressively greater amounts of salsa on his eggs. He obviously liked salsa, and the heat either enhanced that liking or was certainly no barrier to his eating more of that food. He had clearly habituated to that level of hotness.

How does this sort of habituation occur? The nervous system mechanisms that regulate pain perception are quite complex, and even if pain receptors only slowly adapt to a stimulus, there are other levels at which habituation can occur.[25] The endogenous opioid system plays a critical role in the perception and modulation of pain; endogenous opioids are the brain's own analgesic, and their receptors can be exploited by drugs that can be highly addictive. In the short or medium term, endogenous opioids provide a means for us to deal with pain, perhaps by habituating to it. We are less certain about the role of endogenous opioids over the longer term. In one experiment, the habituation response of a group of individuals to a painful stimulus (heat) applied to the arm was charted over the course of eight days of daily stimulation.[26] As expected, the subjects habituated to the pain, reporting lower pain ratings while being able to withstand an increased level of stimulus as the experiment progressed. To test whether or not endogenous opioids were involved in this process, on days one and eight of the study, the researchers gave half of the subjects a drug called naloxone, which blocks the action of opioids; the other half of the subjects were administered a saline solution (in a classic double-blind protocol). The researchers found that naloxone had no effect on the habituation response, suggesting that central nervous system control of pain habituation does not depend on endogenous opioids.

The cognition of pain is further complicated by the fact that emotion plays a huge role in determining how pain is perceived.

Consider the effectiveness of a mother's kiss on her child's small cut—emotional analgesia sometimes works wonders. The brain networks associated with pain perception necessarily include emotion regions of the brain, such as the anterior cingulate cortex. Since emotion has a critical role in determining a person's state of mind, it is clear that perception of pain is also a product of the context in which the pain is experienced. This context is often determined by culture, which in turn influences levels of emotional expression. Many studies have found that ethnicity is a salient variable in the reporting of pain levels.[27]

All of these factors potentially come into play in determining why or why not an individual might eat and enjoy a pain-inducing food, such as hot peppers. The hot pepper is one of the great success stories of Native American agriculture, along with corn, the potato, the tomato, and so on. Archaeological evidence based on starch microfossils indicates that hot peppers were widely consumed in South and Central America as long as 6,000 years ago (spreading later to the Caribbean islands and the Bahamas) and that they, along with corn, were a ubiquitous component of an ancient tropical agricultural package that persisted for thousands of years.[28] The genus *Capsicum*, to which all peppers belong, likely originated in Bolivia. The wild-type pepper fruit is spicy hot, and this discourages herbivores from munching on the seed-bearing fruit. However, birds cannot taste or are unaffected by capsaicin, the substance that provides the heat, so they serve as the vehicles for pepper seed dispersal in the wild.

Humans *are* affected by capsaicin, yet hot peppers were a staple of New World cuisines for thousands of years, and when introduced to the Old World in the sixteenth century, they quickly found a place in cuisines across Africa and Eurasia. Why eat hot peppers? Psychologist Paul Rozin, who pioneered the psychological study of food preferences and disgust, suggests several reasons they may have found their way into these various cuisines.[29] They are a great source of vitamins A and C. The capsaicin activates the gastrointestinal system, increasing salivation and gut motility, making drier foods more palatable. Most critically, hot peppers serve as taste enhancers, especially in a diet where bland foods dominate. Rozin argues that humans seek "variety within a general familiar constraint, or culinary themes and variation." Hot peppers, individually and in combination, are a great tool for the cook looking to create more variety in the context of a diet that may be limited in general or seasonally.

Rozin and his colleagues have looked at how a "hot tooth," so to speak, develops as a child grows up in a traditional Mexican household, where hot peppers are a fundamental part of the diet. Small children, ages two to six years, were exposed to hot peppers in small quantities, increasing over time. The children observed that peppers were valued in the family environment, although they were allowed to refuse the peppers if they did not like them. However, typically by the ages of five to eight, children were already developing a desire to add peppers to their diet. Thus children were introduced to hot food by a combination of mild social pressure and relatively mild spiciness. As in other educational environments, the children were placed in a position where they were benignly coerced to "discover" the benefits of something that they may have initially resisted.

At the level of individual psychology, Rozin offers two explanations for why people might like eating hot foods. These may apply more to understanding the motivations of people from cultures in which hot peppers are less than traditional and who seem to embrace heat for heat's sake. First, there is the "rollercoaster effect," in which a negative experience becomes positive upon increased exposure, once it is learned that it is not really dangerous. Such an experience can also become somewhat boring, which is why some people need to ratchet up the experience with hotter foods or higher roller coasters. Rozin also hypothesizes that the pain from peppers encourages the release of endogenous opioids, and that repeated exposure may lead to increased release of these analgesic chemicals. This would put eating spicy foods on par with getting a sort of cheap "runner's high."

We did not evolve to eat hot peppers, or to be more precise, hot peppers did not evolve so that we, mammalian omnivores, would like to eat them. But they illustrate the power of human omnivory, which is based in large part on learned food choices derived from collectivized and shared memories in a cultural context. The known history of the hot pepper is wholly agricultural, but once upon a time, more than 6,000 years ago, some Native Americans developed a tasted for wild hot pepper. They realized that the painful sensation they experienced when eating peppers was not permanent, and that the peppers' hotness could be made to work for them (or their cuisine). This was a transformative moment in the history of human taste, laying the groundwork for the cultural evolution of a multitude of cuisines based on the potentiating power of pain. It also ultimately provides us with a window into the basic workings of the human mind.

Tasteful Genetics

People have different food preferences: they like the tastes of different things. Some individuals and some peoples cultivate their senses more, others less. Age seems to dull the sense of taste, as do taste-bud-destroying habits such as smoking or drinking too much alcohol. But at a basic biological level, do people really taste things differently? We all use the same basic biological equipment—taste cells, taste buds, and the like—but is there significant biological variation in how this equipment works?

The notion that there are substantial differences in the sense of taste seems logical, since there is variation present in all manner of biological and physiological domains.[30] But variation in tasting is more covert, even compared to other sensory variation. We don't have the equivalent of glasses or hearing aids for poor tasting ability. Some people express genuine surprise when confronted with variation in tasting ability. I know this from the experience of running anthropology lab classes when I was in graduate school. A classic anthropology laboratory exercise is to have the students taste a piece of paper impregnated with phenylthiocarbamide (PTC), a bitter-tasting chemical. Well, it is bitter-tasting to some people; others cannot taste it all. It works well as an exercise demonstrating human variation, because the students can discover for themselves that human variation exists by compiling a tally of tasters and non-tasters of PTC (as if being in a classroom with individuals whose ancestries can be traced to East Asia, South Asia, East Africa, West Africa, Europe, the Americas, and the Pacific islands were not enough to demonstrate human variation). Often the non-tasters even experience a bit of anxiety when they see others react to the bitter PTC yet are unable to taste anything themselves.

The discovery of PTC tasting variation came about in a workplace incident that today "would curl the toes of the most stoic OSHA officer."[31] In 1931, a chemist named Arthur Fox was working in his laboratory, pouring powdered PTC into a bottle. A co-worker complained about the bitter taste of the flying powder, a taste of which Fox himself was completely unaware. Being scientific types, they took turns tasting the powder until they discovered that they really were different in their ability to taste the molecule. They tested a number of others and soon found that people could be divided into PTC tasters and non-tasters. Another researcher, Albert Blakeslee, soon followed up with a genetics study. He found that the ability to taste PTC was transmitted, more or less, as a simple Mendelian trait, with non-tasting being recessive to the dominant tasting. This means that people who can't taste PTC need to carry two non-tasting genes, while tasters can have either two copies of the tasting gene or one of each. Actually, the genetics are somewhat more complicated than this, since PTC tasting ability seems to be a matter of detecting levels of its concentration rather than a simple either/or proposition, but the Mendelian model works well enough for an anthropology lab class.[32] Incidentally, another chemical that is very similar to PTC, propylthiouracil (PROP), is more typically used to test for PTC tasting ability today; unlike PTC, it is used pharmacologically and therefore investigators have some idea of what safe levels of its consumption are. (That OSHA official can uncurl his toes.)

The ability to taste PTC is almost certainly related to the detection of toxins in various plants. As with capsaicin, plants produce these compounds to discourage herbivores from eating them. The chemicals they produce are similar enough to PTC to have the same taste. These compounds are particularly common in the widely cultivated cruciferous vegetables (members of the family Cruciferae), which include cabbage, mustard greens, broccoli, Brussels sprouts—a veritable murderer's row of vegetables for some delicate palates. They are also common in cassava (manioc), a root vegetable that is a staple for millions of tropical peoples. The bitterness has to some extent been bred out of these plants over the thousands of years they have been cultivated, but enough remains to contribute to the plants' unique taste profiles.

The thing that has baffled geneticists for decades is why anyone is a PTC non-taster, since being a taster is the trait that presumably was most valuable in a natural setting. Variation in tasting ability for PTC is what is known as a genetic polymorphism—a trait that shows significant genetic variation in a population. PTC tasting ability has been studied in literally hundreds of populations involving thousands of individuals, and it is polymorphic in almost every one (a single population of Brazilian Indians is the only one known to be all tasters).[33] This does not mean that it does not vary among populations—it does, although strong regional patterns are not obvious. Overall, most people (70–80 percent) are tasters, while a significant minority are non-tasters. The frequency of the non-tasting allele (variant) of the PTC gene is about 45 percent.

Modern molecular genetics has considerably widened our perspective on the classical genetics of PTC tasting. In humans, the protein receptors that determine the ability to detect bitterness are coded by genes from the TAS2R family, of which there are at least twenty-five types; genes from the TAS1R family are responsible for sweet and umami tasting.[34] The PTC tasting receptor is coded specifically by gene TAS2R38. There are several variants of this gene, beyond simply taster and non-taster. PROP tasting does indeed depend on this gene, but there are important differences between PTC and PROP in how the receptor responds to changes in concentration.

The large number of genes involved in tasting bitter substances, plus variation within the TAS2R38 group, may be one reason why it has been difficult to find significant real-world associations with PTC tasting ability.[35] Despite many efforts, no simple correspondence between PTC tasting status and preference for bitter foods has been discovered; laboratory preferences do not always translate into patterns of consumption. Some studies do show an association between PROP tasting ability and reduced vegetable consumption, but others fail to find such an association (and in cultures where vegetable consumption tends to be low, as it is in the United States in general, genetic taste preferences contribute little to the observed variation). All this is not to say that correspondences between tasting status and food preferences do not exist; rather, so many other factors come into play that it is hard to discern a pattern through the noise (which from this perspective could include factors such as cultural food environment).

Natural selection does provide an explanation for the distribution of one receptor for bitterness. This receptor, produced by the gene TAS2R16, is especially adept at detecting cyanogenic glycosides—toxic substances that release cyanide when they are digested in the intestine. Using sophisticated mathematical models based on gene frequency data from sixty populations, Nicole Soranzo and colleagues determined that there was strong natural selection for this gene between approximately 80,000 and 800,000 years ago (admittedly a pretty big time window in the context of later *Homo* evolution).[36] The ability to detect cyanogenic compounds would have been useful and selected for in a hominin species that was both omnivorous and expanding its range. Interestingly, a low-sensitivity variant of this gene (a kind of nontaster version) is still found in high frequency in central African populations. This is an area where cassava is widely consumed. Cassava contains cyanogenic glycosides that must be neutralized or eliminated by processing before it is consumed; cassava may also confer protection against malaria, which is a critical issue in tropical populations around the world. So the polymorphism for TAS2R16 may be maintained due to the fact that non-taster allele is part of a dietary/disease-protection complex in a specific malarial environment.

No such advantage has been found for tasting or not tasting PTC, although there has been much effort devoted to finding a disease connection. Mathematical models suggest that in many populations both tasting and non-tasting variants of the gene are being actively maintained by natural selection.[37] This "balancing selection" occurs when one allele has an advantage up to a certain frequency in a population, but if it exceeds that frequency, the advantage is lost. Stephen Wooding and colleagues hypothesize that the polymorphism for PTC tasting is maintained because the non-taster of PTC is actually the taster of another bitter substance, as yet undiscovered.[38] Although the idea is somewhat undeveloped, I think it is potentially quite interesting. It suggests that humans have found a shortcut to expanding their genetic repertoire of bitter tasting receptors by having one receptor do two jobs. This may have worked together with language, in that our ancestors could share knowledge about toxic plants with other individuals in the culture.

Evidence for significant genetic variation in salt, sweet, and umami tasting ability has yet to be discovered. And just as PTC tasting ability is difficult to link to preferences regarding bitterness, it is difficult to associate it with preferences for other substances. However, among African American women, PTC tasters drink less alcohol and are less likely to be dependent on nicotine than non-tasters.[39] These associations have not been found in European American men or women or in African American men. Again, this indicates that PTC ability plays a role in shaping preferences, but this role is expressed in the context of a host of other genes and the cultural environment.

Although at this point the evidence for extensive variation in taste receptor physiology (other than for bitterness) is somewhat less than overwhelming, there is still justification for Valerie Duffy's claim that "there is normal variation in oral sensation across individuals, ranging from those living in a neon orosensory world to those living in a pastel world."[40] The research of Duffy and her colleagues shows that it is not just the specific activity of receptors that influences taste perception but the number of them as well. Based on PROP tasting ability, the researchers have identified a group of "supertasters" who find that substance intensely bitter rather than just plain old bitter. These supertasters of PROP, who include more women than men, have increased numbers of taste structures (fungiform papillae and taste pores) on the tongue, which is presumably the source of their extra sensitivity. These individuals are also more sensitive to the creaminess of fat and to oral pain.

The supertasters illustrate how other biological dimensions beyond the structure of the taste receptors themselves help to make people different in tasting ability. Genetic differences in odor detection probably play an active role in this phenomenon, for example.[41] So the culture a person lives in and the way he or she grew up tell only part of the story of why any individual eats the way he or she does—the sensory palette encoded in the genes also plays a role.

Why is understanding biological variation in taste perception important, beyond any information it may reveal about our biological origins? One reason is that so many people want to change their diets, whether for health reasons or in the service of beauty. Although taste genetics are just one factor among many, knowledge about our taste genetics could help us modify our diet for the better.

Foodgasm

Food and sex, sex and food. For humans, the two are inextricably linked, and not only because they represent the most basic drives any sexually reproducing organism possessing a nervous system can have. Over the course of human evolution, a sexual division of labor evolved in which males and females cooperated by specializing in how they acquired food to nourish their children.[42] Human offspring, with large, slowly growing, slow-to-train brains, had nutritional needs that outstripped the provisioning ability of a lone mother, especially if she had more than one dependent offspring at a time. In contrast to the great apes, at some point adult male hominins (otherwise known as fathers) became part of the food supply chain for young hominins. In traditional human hunter-gatherer cultures, the sexual division of labor can generally be characterized this way: men provide food in larger, less predictably available packages (e.g., large game animals), while women, with infants and children in tow, specialize in smaller but more predictable food items. How this state of affairs evolved is a matter of much discussion and some controversy, but it illustrates how human sexual relationships (based to some extent on pair bonds) became focused on an economic partnership that involved supplying food for young.

There is no reproduction without sex, of course, and food-providing ability seems to have been part of male-female courtship for a very long time, likely dating back to before our ancestors split off from the great apes. The apes have a good intuitive grasp of the notion of trade and exchange. They understand, for example, that services provided in one context can be exchanged for goods in another, and vice versa. Among highly interactive species such as chimpanzees, dominance hierarchies and stable alliances are based on the dynamics of exchange. Meat, a rare and highly valued food among chimpanzees, can be an important currency in chimpanzee society. As primatologist Craig Stanford writes, "The interplay of meat with sex, political networks, and status display is typical of the strategic meat-sharing pattern seen among chimpanzees. . . . The evolutionary legacy of our hunting and scavenging past lies therefore not so much in the hunt but in the division of spoils."[43] Chimpanzee males (and it is males who do most of the hunting of animals large enough to share) can trade meat for several things, but one of the most important is access to ovulating females. In other words, if a male chimpanzee is willing to share some of his prey with a female chimpanzee, then she is more likely to consent to mate with him.

Chimpanzee males do not have a provisioning role in raising their offspring, so sharing meat with a female is attractive strictly in terms of mating in the here and now. In humans, it is widely acknowledged that food can be an integral part of courtship. The attractiveness of the food-providing ability of a man can be assessed by a woman in both the short term (to help her decide whether to have sex with him) and the long term (to help her judge whether he will be a reliable provider of food for possible offspring). Of course, the cliché is that men are far more interested in short-term mating opportunities than in using food to advertise their child-supporting prowess. M. F. K. Fisher, writing from her own dating experiences in mid-twentieth-century America, describes gourmet bachelors in unambiguous terms: "Their approach to gastronomy is basically sexual, since few of them under seventy-nine will bother to produce a good meal unless it is for a pretty woman. Few of them at any age will consciously ponder on the aphrodisiac qualities of the dishes they serve forth, but subconsciously they use what tricks they have to make their little banquets, whether intimate or merely convivial, lead as subtly as possible to the hoped-for bedding down."[44]

The link between food and sex is both reinforced and revealed in language. Metaphors for sexual acts or anatomy often incorporate the terminology of food and eating, and in many cases the same words or phrases are used to describe both sexual and culinary acts. Like eating, romantic sexual intimacy typically begins with the lips and tongue before the action moves to other, lower areas of the body. Perhaps this anatomical convergence helps to foster the linguistic linkage between the two. Apparently these links are common in nearly all languages. Claude Lévi-Strauss writes, "As is the case the world over, the South American languages bear witness to the fact that the two aspects [food and sex] are closely linked. The Tupari express coitus by locutions whose literal meaning is 'to eat the vagina' *(kümä ka)*, 'to eat the penis' *(ang ka)*. . . . The Caingang dialects of southern Brazil have a verb that means both to 'copulate' and 'eat'; in certain cases it may be necessary to specify 'with the penis' to avoid ambiguity."[45]

In English, there seems to be a directionality in the linguistic or symbolic interchange between food and sex, going mostly in the direction of food to sex. When this directionality is violated, it can be quite surprising or even shocking. The feminist artist Judy Chicago's groundbreaking 1970s piece *The Dinner Party* was shocking and challenging for many reasons, not least of which was the extensive use of vaginal or vulvar imagery served quite literally on dinner plates. And it certainly challenges the conventions of food writing when contrarian restaurateur-philosopher Kenny Shopsin writes, "Bacon pancakes and bacon French toast both remind me of pussy. . . . When you flip the pancakes back to serve them bacon side up, the bacon is *in* there, enveloped by soft walls. It's really very sexy."[46] Shopsin's imagery stands out because we tend to gastronomize the sexual rather than eroticize our food.

There is one major exception to this tendency, and it is why I am discussing sex in a chapter on food, brain, and the senses. It is not at all unusual for someone to describe the sensation upon eating something delicious as "orgasmic." There is even a neologism for this, *foodgasm*, which has been defined several times in the online Urban Dictionary.[47] One definition reads: "An event occurring upon the consumption of some incredibly tasty food. Usually contains many random vocal noises including, but not limited to: moans, sighs, screams of joy, etc. [A]lso, random facial expressions may be included." Another, more succinct definition: "The euphoric sensation upon tasting amazingly delicious food."

Now, even though some people achieve orgasm through unconventional routes (such as during tooth brushing), I don't think there is any claim here that a foodgasm is accompanied in any way, shape, or form by an actual orgasm.[48] But clearly some people feel something quite special, whatever that might be, when they eat something that really hits the spot. Since this phenomenon does not simply occur upon eating something that activates the orbitofrontal region of the cerebral cortex and amygdala, we need to look beyond the tasting brain to figure out how such a sensation might arise. Unfortunately, there do not seem to be any functional brain imaging studies that address this issue. However, people have been able to orgasm under the extraordinary conditions of an imaging study, and the results of those studies might provide us with some insights.

Janniko Georgiadis and colleagues have looked at the brain's activity during orgasm using positron emission tomography (PET) scanning.[49] In their experiments, subjects were manually stimulated by their partners during PET scanning; both males and females were included in the subject group. Brain activation was measured during both genital stimulation and orgasm. Since male and female subjects were scanned under identical conditions, it was possible for the researchers to make direct comparisons between male and female responses to genital stimulation.

Given that women are typically slower to respond to genital stimulation, women were allowed three separate sessions to achieve orgasm, while the men had two.

The results of the studies indicate that men's and women's brains are quite different in their response to genital stimulation. Why this should be the case is somewhat intriguing, in that the penis and clitoris derive embryonically from the same tissue. Since the experimental subjects were not able to see their partners, Georgiadis and colleagues speculate that there may be gender differences in how and to what extent the subjects visualize the activity of their partners.

In contrast to genital stimulation, orgasm prompted very similar brain responses in both men and women. One exception was a small brain stem area (the periaqueductal gray matter) involved in pain suppression, which showed higher activation in men than in women. Otherwise, in both men and women, orgasm activated several parts of the brain stem, some of which reflect the cardiovascular arousal that occurs during orgasm. Of great interest is the fact that orgasm leads to significant deactivation in the orbitofrontal cortex of both men and women. Following the orbitofrontal functional maps developed by Edmund Rolls and others, Georgiadis and his colleagues suggest that deactivation of the lateral orbitofrontal cortex corresponds with sexual disinhibition. In contrast, when men view sexual images but are told to suppress their sexual feelings, the lateral orbitofrontal cortex is activated. The subjects in the orgasm studies were told to delay orgasm until a certain period of time had passed; thus deactivation in this region is thought to reflect the sexual release associated with orgasm.

From the foodgasm perspective, the finding of orgasmic deactivation in the middle orbitofrontal cortex may be the most relevant result of these functional brain studies. This is the part of the orbitofrontal cortex that is associated with satiation and subjective pleasantness of taste.[50] Georgiadis and colleagues do not think that taste has anything to do with it; rather, the decrease in activation during orgasm mirrors the deactivation in this region that occurs with satiety. Orgasm is typically characterized as a process of buildup and release, followed by a feeling of fulfillment that may be subjectively similar to satiation.

At some neural level a foodgasm may indeed be evocative of an orgasm. Satiation is an active inhibitory process typically associated with consuming a certain amount of food to a point where the body unconsciously desires to stop eating. An orgasm similarly signals the termination of a sexual episode. In contrast, a foodgasm typically occurs upon an initial taste, so true satiation cannot be a factor. However, if there is heightened anticipation and tension, then a single extremely pleasant taste may lead to immediate satiation. This unusually rapid satiation may evoke a postorgasmic feeling of sexual satiation. When this feeling is coupled with the hedonic rush of tasting a much anticipated delicious morsel of food, an individual could have a profound psychophysical response to food that goes far beyond simply noting that it tastes good. The orbitofrontal cortex, where orgasm and taste perception overlap, is likely the critical region for the foodgasm. It isn't the same as an orgasm, but it is nothing to sneeze at, either.

Tasting Better

There is nothing wrong with not particularly enjoying the taste of food; it is not a crime to shovel food into your mouth as though you are pumping gas into a car. Not everyone is French, or meant to be a gourmand. All things being equal, bland foods will get a person from the cradle to the grave just as well as spicy or complexly flavored ones. Many people have quite satisfactory food lives even though they cannot experience foodgasm. For us primates, taste exists more to help us identify safe and nutritious foods than for our enjoyment.

Yet for those fortunate to have access to a steady and varied supply of food, it seems a shame to not make the fullest use of our ability to taste. Humans taste food in a much richer and more complex cultural and cognitive environment than does any other animal. Our ability to combine and create flavors, to remember and pass them on across generations, forms the basis of the world's different traditional cuisines. In developed countries, the modern cooking environment grants cooks and diners access to an extraordinary range of tastes and flavors from almost all of the world's cultures. Eaters do not have to settle for the insipid, bland, or monotonous.

All people pick the areas in their lives in which they "settle" rather than make the extra effort to obtain something a bit more special or rewarding. But the sensory pleasures of the table are within such easy reach for many people today that it seems a shame to let them pass by. Tasting food rather than simply eating it means slowing down and giving priority to qualities of foods beyond their ability to fill us up. Except for the most extraordinary cases, taste-blindness can be cured.

4

EATING MORE, EATING LESS

Strange to say, the ability to live on the eucharist, and to resist
starvation by diabolical power, died out with the middle ages,
and was replaced by the "fasting girls," who still continue to
amuse us with their vagaries. To the consideration of some of
the more striking instances of more recent times the attention
of the reader is invited, in the confidence that much of interest
in the study of the "History of Human Folly" will be adduced.

—WILLIAM A. HAMMOND, M.D., *Fasting Girls: Their Physiology
and Pathology* (G. P. Putnam's Sons, 1879)

OVER THE COURSE OF EVOLUTION, from jungles to deserts to
mountains, humans have faced all manner of challenging environments. One might think that an environment in which an
abundance of food and calories is readily available at all times of
the year with minimal physical exertion would not be particularly challenging. Yet in the urbanized, developed (and developing) world, just such an environment is now widely seen as being
quite hazardous. The issue is not survival—making it through
childhood and reaching adulthood is the norm these days—but
long-term health.

The modern food environment has precipitated a "global epidemic of obesity."[1] This epidemic began in the United States and

Western Europe but is now clearly showing its effects in the rapidly developing nations of Latin America, Asia, and the Pacific islands. Although hunger is still a problem throughout much of the world, the issues associated with having too much food are increasingly seen as a ticking time bomb, undermining the health of nations.

For many people living in the evolutionarily unprecedented food environment of the developed world, the abundance of easily available food has contributed to the problem of having an over-abundance of fat on the body. Woody Allen, channeling Dostoyevsky, put his finger on it decades ago: for many people today, the main existential crisis they face is centered on their excess weight: "I am fat. I am disgustingly fat. I am the fattest human I know. I have nothing but excess poundage all over my body. My fingers are fat. My wrists are fat. My eyes are fat. (Can you imagine fat eyes?)"[2]

Weight can dictate a person's quality of life and sense of well-being, not simply in terms of health and mobility but also in relation to social acceptability, professional achievement, and romantic success. Whatever else is going on in a person's life, weight provides a means of assessing it in the light of cold, hard numbers—not just pure poundage but amounts lost or gained, targets set or missed, body mass index, metabolic set point, and so on.

The potential health dangers of obesity are well known. These are not generally acute dangers, except in older individuals whose bodies have in various ways worn out from the cumulative effects of excess weight. In contrast to our knowledge of the health risks of obesity, the causes underlying the epidemic are far from clear. Yes, at the individual level, the modern food environment has made it easy to tip the balance between energy intake and expenditure to the intake side, leading to increased numbers of overweight people. But why exactly this balance has tipped so strongly is still debated. A review by a wide-ranging group of experts in the field concludes that what they call "the big two," commercially driven trends toward overconsumption and institutionally driven reductions in physical activity, are significant but probably insufficient to explain the full scope of the epidemic.[3] Other factors, such as changes in sleep patterns, the uterine environment in which a fetus develops, a child's family environment, and medication usage, may also be contributing to the spread of obesity. Some researchers argue that it is not just the number but also the kind of calories in the modern diet that lead to obesity, pointing specifically to the increased consumption of simple, easily digested carbohydrates.[4]

Whatever the global causes of more people becoming fatter, for the individual who wants to lose weight the problem is an entirely local one. At first glance, the solution to being obese is easy: simply reverse the imbalance between energy intake and expenditure. But it is quite clear that in practice this is not so simple. Complicating things even more is the fact that people vary greatly in metabolism and physiology; a combination of calorie intake and activity level that leads to one person losing weight may result in no weight loss or even weight gain in another. Beyond that, people vary cognitively in their approaches to food and eating, influenced in different ways and to different degrees by the interacting histories—evolutionary, cultural, familial, individual—that go into making a thinking, eating person.

Eating More, Naturally

In Okinawa, home of some of the longest-lived people on earth, there is a traditional expression: *haru hachi-bu*, or "eat until you are 80 percent full."[5] This advice may have served Okinawans particularly well, as some researchers hypothesize that one of the reasons for their unusual longevity may be caloric restriction (which has been shown in many experimental animals to be a reliable way to extend life span). Even if life extension is not the goal, given what we know about the relationships among health, body weight, and longevity, this advice is probably worth taking in order to avoid the long-term consequences of being over-weight. However, many people would probably say that it is easier to overshoot than undershoot by 20 percent when it comes to filling up. Eating too much, at least on occasion, is to many people the most natural thing in the word. In the hunter-gatherer past, and even deeper in our primate evolutionary history, eating more when food is available was undoubtedly adaptive. Food supplies can never be certain, after all.

Our larger-brained ancestors survived and ultimately thrived due in part to their ability to hunt very large game.[6] The amount of meat our ancestors could obtain all in one go far exceeded the amount that primatologists have observed chimpanzees capture on their hunts. Without a means of preservation, however, there would be pressure on the hunters to consume as much as possi-ble in a short time. Even so, in many cases there would be much more meat than they could possibly consume, so there were am-ple opportunities for sharing with kin and community. At some point, sharing among hunting hominins became much more elaborate than the sharing that chimpanzees engage in. Ulti-mately, human food sharing reached the level of the feast—the ritualized, community-based sharing of large quantities of food. Archaeologist Martin Jones traces the origins of feasting to the distribution of meat that developed around large game kills.[7] With the advent of agriculture, the feasting around the hunt transferred to the sharing of the seasonal abundance at harvest. Ultimately, in cultures throughout the world, feasting became one of the core communal activities around which cultural identity and cohesion were established and maintained.

The sense of familial and cultural solidarity that comes with feasting can be very pleasant and reassuring, rewarding in ways both nutritional and psychological. Eating to 120 percent full is probably a pretty standard performance at a feast, so over the course of a million-plus years of feasting, a strong psychological association likely evolved between being extra full and a sense of social well-being. The importance of social life to human beings cannot be underestimated. The common ancestor we shared with chimpanzees millions of years ago was likely a highly social species, just as chimpanzees and we are today. Over the course of our evolution, humans gained a tool—language—that has helped our kind to ratchet up all aspects of social life. Sharing food both on a day-to-day basis and on special occasions is part of the complex of behaviors that now defines how humans are social. In a less profound or obvious way, occasional overeating may also be part of this complex.

High-quality fatty and sweet foods were rare in the traditional hunter-gatherer diet, but they were highly prized. The sweet tooth is well known, but do humans have a "fat tooth" as well? As we discussed earlier, the energy demands of the human brain require that we eat a substantial amount of high-energy food, and especially during its development, the brain may need a steady supply of fatty acids.[8] There is also evidence that even though "fatty" is not a primary taste, humans have dedicated sensory pathways for its detection.[9] We know from accounts of hunter-gatherers and from the archaeological record that the fatty marrow of long bones, for example, was enthusiastically eaten, a bonanza of fat in otherwise lean wild game meat. Beyond the easily accessible marrow of long bones, hunter-gatherers often went in search of fat from the spongy bone of animals (usually the ends of long bones), which does not so easily yield its grease. Archaeologist Alan Outram describes the process of extracting it:

> To do this, you have to break up the bones into little pieces, which is a very labor-intensive job. The fragments then have to be boiled in water to render out the fat, which floats to the surface and can be skimmed after cooling. This may not be difficult in the modern context, but in an early prehistoric context, without metal cauldrons, this had to be carried out in pits, buckets, or pots by heating up rocks and adding them to the water to bring it to the boil. This involves an incredible amount of effort and fuel for a relatively limited amount of fat.[10]

The intense natural desire for fat lays the foundation for overeating fatty foods when they are available in great abundance.

The human fat tooth has great antiquity, although probably not as great as that of the sweet tooth. The sweet-tasting simple carbohydrates available in ripe fruits are an attractive source of calories to most monkeys and apes, and humans fall right in line here.[11] But as countless lab studies have shown, almost any primate would probably prefer to get sweets in purer and more concentrated form than that available in ripe fruits. People who live in developed countries with ready access to essentially unlimited quantities of refined sugars and sugary products eat an astonishing amount of simple carbohydrate. United States Department of Agriculture figures put the annual per capita American consumption of sugar and sweeteners at 131 pounds in 2010, or about 2.5 pounds per week.[12] This is up from "only" 119 pounds in

1970, although down from a peak of 151 pounds in 1999. (Actually, the increase in sugar consumption just about matches the overall increase in food consumption.)

The amount of carbohydrate consumed in the modern diet is roughly similar to that in the Paleolithic diet.[13] However, the majority of carbs we eat today come in the form of simple sugars, which made up only a small percentage of the carbs of the traditional diet. This was clearly not by choice for the traditional hunter-gatherers, however. If concentrated sweet foods were available, they were pursued, sometimes at great risk or time investment (e.g., collecting and processing maple syrup among northeastern Native American groups). The traditional lifeways of the Hadza of East Africa, one of the last traditional subsistence groups left, have been intensely studied for decades. Recently anthropologists Frank Marlowe and Julia Berbesque asked Hadza men and women to rank their preference for five of their most common foods: tubers, berries, meat, fruit of the baobab tree, and honey.[14] Honey was by far the favorite food for both men and women, and tubers, a bland but reliable fallback food, were the least preferred. Men had meat as their second choice, while women picked berries. Men are largely responsible for collecting honey, which is found in hives high up in tall baobab trees. The risk of injury or death from falling is real and considerable, but the honey is a desired enough product to be worth the effort. Hadza men typically go alone to hunt or forage for honey. They take back to camp about 90 percent of the meat they hunt, but only about 50 percent of the honey they collect.

The evolved preferences for fatty and sweet, and for salty as well, establish a foundation for occasional overindulgence under traditional conditions.[15] However, in the modern environment of food plenty, these preferences, combined with the social and psychological rewards we associate with feasting, may contribute to the phenomenon of too many people eating too much too often. Whatever its specific causes, the obesity epidemic that is occurring in developed countries throughout the world is ultimately a result of placing bodies and minds evolved for one environment in one that is wholly different. Paradoxically, this environment that makes calories so accessible puts much lower caloric demands on the people living in it. As Michael Power and Jay Schulkin write: "One of the asymmetries that likely contribute to overweight and obesity is that motivation to indulge in food can be greater than motivation to indulge in physical activity."[16] In the past, obtaining food required considerable physical activity, something that is far from true today.

If eating more is a natural reaction to an unnatural environment, what about eating less? There are always constraints on the amount of time and energy that can be devoted to eating. Any social primate must also worry about finding a mate, avoiding being eaten, and contributing to the maintenance and protection of the social group. These activities help to fill the days of a typical monkey or ape, limiting the amount of time the animal can spend on gathering and eating—but they say nothing about the amount eaten. Although most primates may naturally gain or lose weight with fluctuations in the quantity and quality of the food supply, generally there is no time when they eat less in a strategic or adaptive way, regardless of the food available. Captive primates living in zoos often fight the battle of the bulge if their diets and activity are not closely monitored.[17]

Other animals do strategically not eat at certain times.[18] Hibernating mammals go for long periods without eating, several bird species stop eating while incubating their eggs or protecting their young, male red deer and elephant seals do not eat while they are defending their harems of females during mating season, and many species eat very little during migration. Animal anorexias are normal, adaptive responses to the natural and social environment in which these animals live.

For humans, there seems to be little evolutionary basis for losing weight or eating less when food is available. That is perhaps one important factor, among many, that makes losing weight so hard. It also provides the backdrop for why deliberately not eating is such a culturally powerful statement. The ritual of fasting is as significant a part of many cultures as feasting is. Culturally prescribed fasting is a demonstration of dedication, willpower, or penitence that signals to the community a willingness to do something that comes very unnaturally—to make a true, if transient, sacrifice. The medieval "fasting girls" referred to in the quote that opens this chapter took this sacrifice to an extreme. It was "folly" indeed to believe that they could go years without food, but that they demonstrated their dedication to God and Christ by denying sustenance to the point of near starvation could not be denied. In the nineteenth century, as religious fashions changed, fasting girls and women began the transition from "sainthood to patienthood," as Joan Jacobs Brumberg describes it.[19] It remains a matter of debate as to whether the *anorexia mirabilis* of the Middle Ages is the same condition as the present-day psychiatric disorder anorexia nervosa, but at least in terms of their effects on the body they are certainly very similar.

In the modern world, the choices people make to eat more or less, too much or too little, are influenced by a great range of factors. The cognition of eating, which in many ways reflects ancient rewards and imperatives, intersects with other cognitive processes shaped by the novel cultural and nutritional environment of the developed, urbanized world. When food is always available, people discover that eating can have powerful effects on behavior even when they are not hungry. Eating when sad or happy, bored or busy, or simply because a desirable food is there can be psychologically satisfying and habit-forming.[20] Conversely, in a world in which eating less is often regarded as a laudable goal, it is easy to see how some individuals may come to consider eating virtually nothing at all to be the ultimate achievement in self-control.

Ingesting with the Brain

In mammals, the brain is the ultimate decider in terms of eating— when, where, how much, and how often. Of course, eating control varies considerably among species. In general, the greater the amount of higher-level cognitive processing power a species has, the greater the number of factors, aside from hunger and food accessibility, that will come into play. However, as in so many complex behavioral patterns, it is folly to think that higher-level processing supersedes what goes on at lower levels. There is, to use a rather jargony phrase, vertical integration.

The vertical integration that monitors and regulates eating behavior and appetite extends from the top of brain to the depths of the gut.[21] It is a complex mechanism, the details of which are not well understood. We do know that it involves all of the organs of digestion, which in turn communicate with the brain via the peripheral nerves, a variety of neuron signaling molecules that operate in both the brain and the gut, and hormones (e.g., insulin). A structure in the brain called the hypothalamus, located in the center of the brain at the base of the cerebral hemispheres, is critical in the regulation of appetite and mediating the communication between the mind and viscera. Different parts of the hypothalamus are dedicated to satiety and feeding. Laboratory animals who have the satiety-regulating part of the hypothalamus destroyed overeat and get fat, while those whose feeding part is lesioned stop eating altogether. The hypothalamus is not simply the feeding junction of the brain and body, however; it helps to regulate all manner of autonomic processes in the body, such as temperature, sleep, sexual behavior, and highly emotion-laden behaviors such as aggression.

Once food gets past the sensory gatekeepers in the mouth, it enters the gastrointestinal tract, which includes the stomach and large and small intestines.[22] These organs are not just passive vessels that hold food while its nutrients are extracted, but active players in regulating the whole ingestive process. Nerve fibers line the tract, monitoring the volume and composition of its contents and relaying that information to the brain stem and then to the hypothalamus. The nutrient content of the food in the gut is assessed by cells lining the tract; they convey information to areas of the brain using specialized molecules that circulate in the bloodstream. Postingestive organs, such as the liver and pancreas, are also in communication with the brain. For example, the pancreas is sensitive to levels of glucose in the bloodstream, secreting insulin as those levels rise; insulin, in turn, acts directly on the hypothalamus and other regions of the brain (see Chapter 5).

To get a sense of the complexity of appetite regulation, let us consider the role and actions of just one of the hormones involved in this system. Leptin is a hormone produced by the fat tissues of the body, and receptors for this hormone are found in high density in the hypothalamus. When leptin was originally discovered in the early 1990s, researchers thought its role was to signal the brain to stop eating. This was a logical conclusion, since obese strains of mice with defective leptin-producing genes had long been used in laboratory research. When these mice were given leptin, they lost weight. Leptin appeared to be a potential silver bullet in the fight against obesity.[23]

Unfortunately, leptin's promise has not been easy to fulfill. Leptin is produced in direct correlation to the amount of body fat present; thus obese people already produce large amounts of this hormone. This suggests that the simplest treatment option, leptin supplements, would have to further increase the already high levels of leptin in the bodies of obese people. Still, given the potential profit in producing an obesity pill, clinical trials with a recombinant form of leptin were undertaken.[24] The results were mixed. Dieting subjects lost more weight with leptin than with a placebo, so that was a positive; however, leptin induced wildly varying responses in the dieting subjects. Some lost large amounts of weight, while others actually gained weight. There may be a subset of obese subjects who do not produce much leptin, and they might be helped by supplementation. However, for most obese people, some of whom may be obese because they are relatively resistant to the activity of leptin, taking more of the hormone would do no good.

We now recognize that leptin is not an eating suppression hormone but one that signals to the brain that energy reserves, in the form of fat, are plentiful. When leptin levels are low, this encourages an animal to either eat more or conserve energy. Being too fat is such a rare occurrence in nature that a normal physiological function for high leptin levels has never really been subject to the forces of natural selection.[25] It is no surprise, then, that high levels of leptin, whether produced by the body or delivered from an external source, do not provoke a standard, clinically useful response.

The scientific and public response to leptin as a potential wonder drug illustrates how the developed world's topsy-turvy nutritional environment of caloric excess clouds perspectives on how bodies evolved and how they function. A similar situation occurred in the 1960s, when it was discovered that some people stop producing the enzyme to digest lactose, the sugar found in milk, when they become adults.[26] It is now well known that the vast majority of adults in the world cannot digest lactose (that is, they are lactose intolerant); this is the normal mammalian condition. Only in populations where there has been a long history of dairying and fresh milk use has there been natural selection for people who can digest milk in adulthood. However, the first scientists to study lactose digestion were predominantly descendants of northern European dairying cultures, and so they considered lactose intolerance in adulthood to be a genetic disease and not the normal human condition. Given their environment, perhaps it was an understandable mistake, but they were definitely wrong. For leptin, discovered in an environment in which great emphasis is placed on finding a quick fix for obesity, it is understandable that it was first interpreted and promoted as an appetite suppressant. There is no doubt that leptin is a modulator of appetite (one of many substances that have a part in this process), but its natural role may be expressed when it is at lower levels in the body, rather than in the high levels seen with obesity.

So is leptin of no clinical value in the treatment of obesity? Not necessarily. Some studies indicate that leptin used in conjunction with other regulatory hormones may be more effective than leptin alone for weight loss.[27] Perhaps someday there will be a hormonal cocktail to control obesity, rather than a silver bullet. A different, perhaps more subtle approach would deploy leptin to address the problem of yo-yo weight loss—to break the cycle of loss and then gain that plagues so many dieters. Leptin levels decline with weight loss, which, as we discussed above, leads to suppressed energy expenditure and increased appetite and hunger. These, of course, promote the regaining of weight recently lost.

Michael Rosenbaum and his colleagues have found that replacement doses of leptin given to people who have lost 10 percent of their body weight can help them keep the weight off by reducing the effects of low leptin levels.[28] Using functional MRI, they have also found that both weight loss and leptin treatment affect the brain activity of dieting individuals looking at food items (which were compared with individuals looking at non-food items).[29] With weight loss, Rosenbaum and colleagues write, "there appear to be increases in activity . . . in systems relating to the emotional, executive, and sensory responses to food while there are decreases in systems involving emotional and cognitive control of food intake."[30] In other words, following weight loss, dieters are in a cognitively vulnerable position—they want to eat more, and at the same time they are less able to control their emotional responses to food—a bad combination for a dieter. Leptin (but not placebo) given during this weight maintenance phase returned brain activity patterns to those seen in the pre-diet condition. Thus at a cognitive level, maintenance dieters on leptin were at less risk for regaining the weight they had lost.

Leptin is just one link in the chain of appetite regulation in the human gut-mind system. It clearly has value, or will have, for some people trying to lose weight and keep the weight off. The basic premise of dieting is fairly simple: eat fewer calories than the amount expended. But this simple formula, which runs counter to the evolutionary, cultural, and (in many cases) personal histories that have shaped our bodies, can be difficult to follow. Bodies do not particularly want to lose weight, metabolically adjusting to decreased caloric intake. Leptin and other links in the appetite regulation chain may someday provide the little extra push that many people seem to need in order to have a healthier future.

Brain Structure and Body Fat

The revolution in neuroimaging allows us to compare the brain structures of all kinds of groups of people—men versus women, old versus young, musicians versus non-musicians, deaf versus hearing, fat versus thin, and so on. Sometimes these comparisons are made in the interest of basic science, to understand human brain variation and its origins at a fundamental level. Some comparisons yield insights into the function of the brain and how it changes with learning and the development of expertise. Research comparing the brains of obese and lean people is driven, unsurprisingly, by concerns about health.

The damage that obesity can do to the body is well known. Heart disease, diabetes, and wear and tear on the joints are some of the most common ailments associated with carrying too much weight. Less appreciated is that obesity poses a risk to the brain, especially the aging brain. Stroke risk rises in association with various forms of heart disease, and cerebrovascular disease itself can cause cognitive impairment and vascular dementia (by restricting local blood flow within the brain). But obesity and type 2 diabetes may more frequently have an impact on cognitive health by hastening the development of the most common form of dementia, Alzheimer's disease.[31]

As we get older, our brains shrink.[32] This is generally referred to as cerebral atrophy, although other parts of the brain besides the cerebral hemispheres shrink with age as well. The gray matter volume of the brain (made up of the neurons) tends to decline slowly across the life span; in contrast, the white matter does not decline through most of adulthood and can even increase in size. But starting around the age of sixty, white matter volume begins to decline rapidly, and overall atrophy of the brain accelerates due to the combined effects of gray matter and white matter decline. This normal pattern of cerebral atrophy is accelerated in Alzheimer's disease and can even predate the development of observable cognitive impairment. Larger brain size may delay the onset of the cognitive and behavioral effects of Alzheimer's, presumably by providing more reserve to buffer the advance of cerebral atrophy.[33] Conversely, anything that accelerates cerebral atrophy may increase the risk of developing Alzheimer's disease or hasten the appearance of symptoms.

An increasing number of structural neuroimaging studies suggest that obesity, or even simply being overweight, may have a negative impact on brain health. Among middle-aged people, increased body mass index (BMI) is associated with reduced total brain volume and increases in neuron and white matter abnormalities, especially in the frontal lobe.[34] Similar associations are also found in studies of older individuals; for example, a study of older Swedish women found a correlation between BMI and increased atrophy in the temporal lobe.[35] This is potentially significant for Alzheimer's disease because the temporal lobe contains some of the regions most critical for memory. It is important to note that the positive correlation between BMI and temporal lobe atrophy was seen even though the majority of women were not obese. A study from Japan looking at a large sample of men and women of varying ages found changes in several brain regions that were correlated with BMI in men (although none in women).[36]

Correlation is one thing, but a more focused result can be obtained by directly comparing brain structure in groups of obese and lean individuals. In one such study, conducted by Nicola Pannacciulli and colleagues, several differences in the brain were found between the two groups: regions in the frontal and parietal lobes, as well as the putamen (one of the nuclei of brain cells buried within the cerebral hemispheres), showed a reduction in gray matter in obese people compared to lean individuals.[37] The investigators pointed out that these regions are all involved in the regulation of taste, reward, and eating behavior. In another study, this one comparing older obese and lean individuals, Paul Thompson and his colleagues found that over a five-year period, obese individuals showed greater atrophy in the frontal lobes, hippocampus, anterior cingulate gyrus, and thalamus, compared to lean individuals.[38] These studies are not designed to establish whether the various brain changes are a result of obesity or a predisposing factor, although the prevailing opinion, based on the growing literature, is that obesity is likely a causative factor.

One established risk factor for developing obesity is possession of a particular allele of the fat mass and obesity-associated (FTO) gene.[39] This variant of the gene is found in 46 percent of Western Europeans and is associated with increased waist circumference and weight. In a study by Paul Thompson's group that involved 206 healthy elderly individuals, possession of this allele was associated with an 8 percent reduction in the size of the frontal lobes and a 12 percent reduction in the occipital lobe. Both of these lobes were also smaller in people with higher BMI. In addition, higher BMI was associated with brain deficits in the temporal and parietal lobes, brain stem, and cerebellum. These correlations held even after other common health indicators, such as cholesterol level and hypertension, were controlled for. In another study by Thompson's group, it was found that in patients diagnosed with Alzheimer's disease and mild cognitive impairment, higher BMI was again correlated with brain volume deficits throughout all of the major lobes of the brain.[40] So in both cognitively intact and clinically impaired elderly individuals, increased BMI is correlated strongly with decreased brain volume.

Why should obesity cause our brains to shrink even more quickly with age than they would otherwise? Damage to blood vessels in the brain associated with diabetes and prediabetic conditions could play a role. Reduced blood flow can directly cause tissue damage to the brain, and it also slows down the clearance of neurotoxic substances that are produced during the development of Alzheimer's disease. In addition, obese individuals are likely to be less physically active than non-obese individuals. Physical activity, which enhances cardiovascular health, is well known to decrease the risk for developing dementia. So elderly obese individuals compound their risk for developing dementia via the combined effects of vascular damage from diabetes-related conditions and the poor maintenance of cerebrovascular health due to inactivity.[41] This all sounds pretty dismal; what's worse, there is ultimately no way to avoid the cumulative effects of brain aging.[42] However, at least diet and exercise provide a fairly straightforward path toward maintaining the health of not only the body but the brain over the life span.

Brain Function and Body Fat

Structural neuroimaging has been useful in demonstrating that obesity has negative effects on brain health. Functional neuroimaging, which scientists use to identify brain networks involved with eating and appetite, is being used to see if people who are overweight or who have eating disorders cognitively process food and appetite signals and stimuli differently than people who do not have these issues. As I have already discussed, the neural networks dedicated to food and eating can be quite complex, involving multiple brain regions at all sorts of different cognitive processing levels. In doing functional neuroimaging studies, researchers have to correlate mental performance on a task to some biological or psychological measure that varies between subject groups or within a group. Although obese versus non-obese is a natural comparison to make, sometimes it is more fruitful to correlate brain function with a continuous variable (e.g., BMI), rather than comparing two distinct groups that are somewhat arbitrarily defined. The task subjects undertake during these studies usually involves looking at pictures of food—say, before or after eating something, or while fasting or full.

Because eating is such a multifaceted behavior, there is potentially a large number of factors that can influence why one person mentally processes food differently from another. The consequences of being obese are relatively universal compared to the psychological variables underlying overeating (not to mention the physiological factors important in weight gain or loss). So when we look at the results from functional neuroimaging studies, it is important to keep in mind that there will be a lot of variability to deal with, some of which is physiological, some of which is cognitive, and some of which results from the different ways researchers set up their tasks and the composition of the groups being studied. Researchers are adept at producing neuroimaging data, which ultimately may make scientific sense only after much replication and statistical meta-analysis. This is a general issue with functional neuroimaging studies, not just those dealing with eating and appetite.

With those caveats in mind, let's look at a few functional imaging studies that serve to make the connection between brain function and potentially important eating-behavior-related variables. One of these variables is gender. Although being overweight or obese is not strongly gendered, in developed countries women are certainly much more likely than men to develop eating disorders.[43] As I will discuss below, the neuroimaging of eating disorders does reveal both structural and functional differences between sufferers and healthy comparison subjects. But is there a gender difference in food stimulus processing in healthy men and women that suggests why women might be more susceptible to eating disorders?

One study points to a potential sex difference in the ability to inhibit brain activity in response to food stimuli. Gene-Jack Wang and his colleagues looked at brain responses while subjects viewed, smelled, and tasted their favorite foods.[44] In one experiment, the subjects did this as normally as possible (well, at least as normally as is possible in a PET scanner), freely experiencing and enjoying the food stimuli. In the contrasting experiment— the cognitive inhibition condition—the subjects were told to inhibit their desire for food and suppress their feelings of hunger (all subjects entered the experiment after fasting). During the voluntary cognitive inhibition condition, the male subjects showed reductions in activation of the amygdala, hippocampus, insula, orbitofrontal cortex, and some of the basal ganglia—all parts of the brain involved in emotional regulation, conditioning, and motivation. The female subjects showed no brain deactivations during cognitive inhibition. Wang and colleagues interpreted this result to mean that women are less able than men to cognitively inhibit their desire for food, which could explain why women tend to have less success in dieting than men.[45] While this is a possible interpretation, I think that more generally the results demonstrate that women and men potentially differ in their motivations concerning food and eating, which could be relevant to gender differences in dieting and eating disorders.

Another variable that could influence mental approaches to eating is personality type. Luca Passamonti and colleagues used functional MRI to see if there was any relationship between performance on a test of external food sensitivity (EFS) and the mental processing of images of bland foods and appetizing foods.[46] External food sensitivity basically describes how responsive an individual is to the external cues of a food (e.g., its appearance) and the extent to which an appealing food makes that person want to eat it.

Not surprisingly, overeating is associated with people who score high on tests of EFS: people with high EFS will say they want to eat a food even if they are not hungry. Passamonti and colleagues scanned their subjects while they were hungry and then again when they were sated. Upon viewing appetizing foods (compared to bland foods), all the subjects activated a network of brain structures (including parts of the basal ganglia, the amygdala, and parts of the frontal lobe) that was predictable based on animal models of feeding. There was no difference here between the high-EFS and low-EFS subjects. However, the researchers also assessed the strength of connectivity between the various activated regions of the network, and calculated the difference in the strength of connectivity between the regions when viewing appetizing and bland foods. Passamonti and colleagues found that connectivity among the regions changes depending on whether the subject is viewing appetizing or bland food. However, the connectivity of higher-EFS individuals changes less than for lower-EFS individuals. This suggests that the network of the higher-EFS subjects is less responsive to cues about the palat-ability of food compared to low-EFS subjects, which may make them less discriminating about what they find appealing. Perhaps of critical importance is that this pattern held whether or not the subjects were hungry. The high-EFS eaters have a less responsive brain network for assessing the palatability of food whether or not they need to eat. Passamonti and colleagues speculate that this less-efficient network may be the reason that high-EFS individu-als are particularly vulnerable to overeating and developing eat-ing disorders.

Governments attempting to address the obesity epidemics in their countries are likely to encourage an epidemic of dieting. One of the most intriguing aspects of variability in eating behavior concerns dieters and dieting: while many people can lose weight, only a relatively small minority can keep it off over the long term. Successful weight loss maintainers (SWLs), defined as dieters who have lost at least 13 kg (30 lbs) and kept it off for at least a year, are such an interesting group that there is even a national registry in the United States to help researchers find them. Compared to normal-weight people, SWLs monitor their diet and weight (and activity) with much greater diligence. They are not successful di-eters by accident. But observing what they do does not tell us much about how they think about food.

A fascinating study by Jeanne McCaffrey and colleagues suggests that SWLs really are different from normal-weight and obese individuals, at least in how they process images of food items.[47] Using functional MRI, McCaffrey and colleagues pre-sented neutral pictures, pictures of low-energy foods (e.g., grains, vegetables, salads), and pictures of high-energy foods (cheese-burgers, french fries, cake) to SWLs, normal-weight people, and obese individuals. Compared to the other two groups, the SWLs showed greater activation in parts of the frontal and temporal lobes when they viewed food items. McCaffrey and colleagues note that the parts of the frontal lobe that were more strongly activated in SWLs were regions associated with the conscious control of behavior—the executive region. When the successful weight loss maintainers look at food, they can consciously inhibit other cues—social, cognitive, emotional—that might otherwise encourage them to eat. This strategy is manifest in the activation of the frontal lobe when they look at pictures of food.

The increased temporal lobe activation in the SWLs occurred in regions important for visual processing. This difference was seen only for food items; there was no difference among the groups when they were looking at neutral items. McCaffrey and colleagues suggest that this increased activity in the visual processing regions may reflect the increased monitoring of food that SWLs apply to keep the weight off. The more closely they pay attention to food, the less likely they are to engage in passive eating.

There was one other significant result from this study. Compared to the normal-weight and SWL subjects, obese subjects showed increased activation in the motor cortex of the frontal lobe when they looked at the pictures of food. The non-obese subjects showed very little activation at all in this region, so it was not just a matter of the obese subjects showing an increased activation; rather, it was an altogether different pattern. McCaffrey and colleagues hypothesize that this activation of the motor region could reflect greater motor readiness on the part of obese subjects in response to food cues.

These three functional neuroimaging studies provide just a small introduction to a future in which researchers will chart with increasing detail the mental phenomenon of eating and how different cognitive approaches to food are associated with different ways of eating. But how, aside from basic knowledge about cognition, can this kind of information help people lose weight? One thing that is certain is that functional neuroimaging will identify numerous ways in which food- and eating-related cognition varies, making it very difficult to implement this knowledge in a general way to help individual dieters.

There is a similar situation developing in the field of medical genetics. Geneticists keep finding more and more genes that are associated with various conditions, diseases, and risks for diseases at the population level, but this type of information is not all that useful for individuals. Nonetheless, advocates for incorporating these data into everyday clinical practice envision the development of something they call "personalized medicine."[48] They foresee a time when the one-size-fits-all medical model will be replaced with one that makes full use of the modern array of clinical tools, including genetic risk factors. A hurdle for the adoption of personalized medicine is that some of the information that comes into play is quite sensitive or invasive by traditional clinical standards. In the past, we had only the vaguest notion of the probabilities of developing certain diseases. Today, however, some genetic studies (such as those that identify the genes associated with breast cancer) can give us much more accurate information about those probabilities. In some cases, clear clinical paths are laid out for individuals likely to develop a disease; other times, there is little that can be done. (Additional controversial issues involve ownership and security of this kind of data. There are also ramifications on many fronts, from the financial to the psychological.)

Neuroimaging is another untraditional invasive technology. No one argues about getting a CT scan to see if you have had a stroke, but what if a functional imaging study shows that your brain processes information about food in a way that is consistent with that of people who overeat as a means of dealing with the emotional trauma of being sexually abused as a child?[49] We are not to that point yet, but the field progresses faster than we can usually anticipate. When we look beyond such potential exposures, it seems likely that if neuroimaging research is going to help individual people lose weight, it will have to be incorporated into a program of personalized medicine. Changing how we eat food means changing how we think it.

At first glance, the results of these imaging studies of how people mentally process food tend to reinforce what is known from more conventional psychological tests or observations. Dieters who keep the weight off show greater activation of regions of the brain associated with high-level control and visual monitoring, as expected. Obese people are more ready to take action when they see food; again, not a surprise. Dieting means change, and while changes in the body can easily be measured objectively, changes in the mind—to attitudes, emotions, processing networks, and so on—are not so easily measured. It's an empirical fact that people can have great temporary success changing their diets and physical habits, only to fall back again and again on old patterns of behavior and eating. Perhaps neuroimaging provides a way to track dieting success at a different level.

David Kessler argues forcefully that an important factor in the developed world's obesity epidemic is something he calls "conditioned hypereating."[50] In this environment of endless nutritional plenty, many people focus only on the psychological rewards of eating (especially eating tasty high-fat and sweet foods), ignoring the brain's signals to stop eating. They become conditioned to eating whether hungry or not, and then to eating more and more.

Dieting cannot simply consist of eating less; rather, it requires a structured eating environment in which the conditioning is initially recognized and then refocused. For people who are conditioned hypereaters, monitoring this deconditioning could be facilitated by functional neuroimaging used as a form of biofeedback.

Besides conditioned hypereating, there are probably many other eating styles that make it easy for people to eat more and difficult to eat less. A personalized neuroimaging profile could help dieters identify their eating styles and monitor their success in changing the underlying psychology that helped them become obese in the first place. This is a ways off, but it was not that long ago that structural neuroimaging was unimaginable, either, and now it is routine. The health consequences and associated financial costs of obesity are so great that any potentially useful tool to attack it, such as functional neuroimaging, should be explored and developed.

Addicted to Food (or Eating)

We need to eat every day. When we are hungry, we crave food. Sometimes we crave specific foods. For example, many people experience intense cravings for salt. This desire makes physiological sense in hot, humid environments where sources of sodium are limited. It is very likely that the taste and desire for salt were shaped during our evolutionary past as tropical primates.[51] However, many people today live in environments that are neither humid nor short of salt. Nonetheless, the brain's reward mechanisms for salt are still designed for the ancestral, low-salt environment. In salt-abundant environments, physiological need is decoupled from psychological reward, and it becomes all too easy for some people to consume too much salt. This has negative health consequences for people whose blood pressure is highly sensitive to salt intake.

When people experience cravings directed at specific foods or flavors, relieving these cravings by eating can bring intense feelings of pleasure; not relieving them makes us feel bad. We are dependent on food, we need it to live, but does it make sense to say we are addicted to food? We could just as well say that we are addicted to air. But the notion of food addiction has in recent years taken on some credibility. By definition, addicts are psychologically different from non-addicts, whatever the drug of choice. Their levels of craving and dependence exceed normal levels, and they are unsatisfied with or unresponsive to drug doses that would satisfy or even incapacitate non-addicts. Drug addicts crave drugs even when they have gotten past the point of enjoying their effects.

The term *food addict* is difficult to define. In drug addiction, external observers can generally identify the negative consequences of continued usage of a drug versus the positive effects of discontinuing it. There are also measurable physiological effects that go along with different forms of drug addiction. But food is different from an addictive drug. Food consumption is fundamentally positive, and continued consumption is mandatory. Most researchers agree that it is very difficult to define a healthy, normal-weight individual as being a food addict, even if he or she eats large quantities of food in general or one food in particular. An exception to this would be in cases of binge eating/bulimia, where a normal weight is maintained by unhealthful means. Michael Lutter and Eric Nestler suggest that at this stage of our understanding, a simple, relatively uncomplicated definition of food addiction may work best: "a loss of control over food intake."[52] Although somewhat vague, this definition acknowledges that there is a potential overlap between food and drug addiction without necessarily equating the two.

How, then, do food and drug addiction converge? Aspects of brain chemistry and functional anatomy appear to point to a common biological basis for these diverse addictions. Overlaps between the effects on the brain of drugs and food have been noted in studies of the brain's endogenous opioids and cannabinoids.[53] The brain pathway that may be most relevant for food addiction involves the neurotransmitter dopamine, which has long been implicated in the regulation of reward, pleasure, and motivation.[54] Many drugs, such as cocaine, activate these natural pleasure and reward centers in the brain; if the drugs are abused, the receptors become less sensitive. The pleasure of eating—and hence the motivation to find food—is also at least partly regulated by dopamine.

Several studies on rodents indicate that dopamine pathways are involved in how mammals process the pleasure derived from food and eating.[55] Even in strains of mice that cannot taste sweet flavors, dopamine is released after the mice feed on a high-energy sweet solution, suggesting that the reward pathway responds to the caloric content of a food, independent of its taste. Obese rats show a desensitization of a certain class of dopamine receptors, similar to the downregulation of dopamine sensitivity seen in human drug addicts. Conversely, rats who have this class of dopamine receptors knocked out quickly develop compulsive food eating behaviors and become obese. These dopamine receptors are also less available in obese humans compared to those who are not obese.[56] Overeating may be a way to compensate for the reduced activity of these dopamine receptors in obese people, although it is also possible that the receptors become less available as a consequence of overeating.

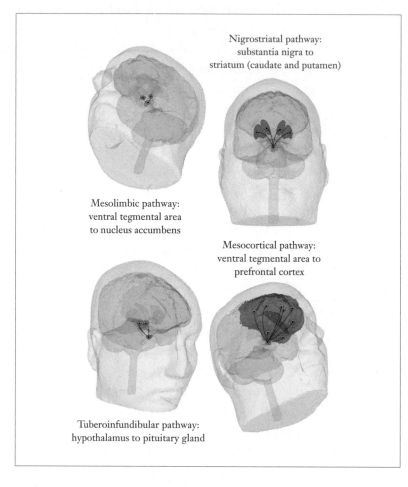

Nigrostriatal pathway:
substantia nigra to
striatum (caudate and putamen)

Mesolimbic pathway:
ventral tegmental area
to nucleus accumbens

Mesocortical pathway:
ventral tegmental area to
prefrontal cortex

Tuberoinfundibular pathway:
hypothalamus to pituitary gland

The neurotransmitter dopamine is used in several brain pathways. The mesolimbic and mesocortical pathways are both important in reward, motivation, and addiction. Both originate in the ventral tegmental area of the brainstem. The mesolimbic pathway connects to the nucleus accumbens, one of the basal ganglia with strong connections to the limbic system. The mesocortical pathway connects to many locations in the prefrontal cortex. The nigrostriatal pathway, which again involves the basal ganglia, is important in the control of movement, while the tuberoinfundibular pathway is important for many processes involving hormonal regulation.

In either case, dopamine is implicated in the reward pathway of eating in humans, and differences in this pathway are evident between non-obese and obese individuals. This is very similar to the situation observed in drug addiction. Susceptibility to developing drug addiction is also likely linked to genetic differences in the dopamine system. Julia Mennella and colleagues looked at whether having an increased preference for sweet tastes is associated with an increased risk for developing alcoholism.[57] They examined children who came from families with and without histories of alcoholism and depressive symptomatology, and tested them to measure their preferred concentration of a sucrose solution. A developmental universal is that all children like sweet tastes, and at higher levels than adults. Eating sweet things also makes children feel better. Mennella and colleagues found that children with a family history of alcoholism and depression had a preference for significantly higher sucrose concentration levels than children who did not have such family histories. For these children with alcoholic and depressive family histories, their preferred level of sucrose concentration was about double the level of sweetness found in a full-calorie cola.

Mennella and her colleagues offered three possible (and not mutually exclusive) explanations for why the children with family histories of depression and alcoholism might prefer more intense sweetness. First, they may have learned from their mothers, who were more likely to be obese and have mood disturbances, to prefer sweeter foods from an early age. In general, obese individuals prefer higher levels of sweetness than do non-obese people. Second, these children may be genetically less sensitive to sweet taste and therefore require more of it to be satisfied. And third, they may require higher levels of sweetness to activate the dopamine reward system, perhaps to compensate for a tendency toward depressed mood. Any of these factors, either alone or in combination, would increase the risk for developing obesity in both childhood and adulthood.

Multiple cognitive factors clearly play a role in the development of obesity. In addition to the level of stimulus required to obtain a cognitive reward, there may be differences in dopamine-related motivation and anticipation for food-related stimuli. In a functional MRI study comparing lean and obese teenage girls, Eric Stice and colleagues looked at the separate effects of anticipation and reward surrounding the consumption of a chocolate milkshake.[58] When anticipating drinking the milkshake, obese girls showed greater activation in the gustatory cortex (the taste-processing regions of the insula and the surrounding frontal lobe) and in parts of the primary body sensory areas. Stice and colleagues interpret these results to mean that the obese girls demonstrated heightened anticipation for the milkshake compared to the lean girls. Conversely, after consuming the milkshake, obese girls had reduced activation in subcortical areas rich in dopamine receptors, suggesting a muted response in the brain's reward pathways. Again, this could be due to either an intrinsically reduced level of dopamine receptor activity or receptor downregulation from a history of overeating.

Further research by Stice and his colleagues has shown that there is a complex interaction within an individual between dopamine genetics and reward activation in response to food, at least in terms of predicting future weight gain.[59] This comes as no surprise; although we focus on the dopamine system in terms of food addiction, many variables are obviously important in overeating. Food reward and anticipation stimulate a much wider network of brain regions in obese individuals than they do in lean ones, suggesting the critical involvement of pathways that do not involve dopamine. The environment can shape not only food availability and preferences but also the critical psychological and cultural associations with food that may influence overeating.

The concept of food addiction is a simplification of eating behavior, and it cannot explain all of the underlying factors and motivations related to overeating and obesity. It is very likely that it explains some of them, however, and that is much better than nothing. And some people may be more likely to be successful at losing weight if they consider themselves to be food addicts and undertake a structured program to help them deal with that behavior (this is much like the way some drug treatment programs work). On the other hand, the phenomenon of hedonic eating is very different from hedonic drug use. Natural selection has shaped the cognition of humans and other animals to derive pleasure from consuming foods when hungry. Hedonic drug use has no such evolutionary foundation.[60] Thus from the perspective of reward and motivation mechanisms in the brain, which evolved in response to the need for food and reproduction, among other things, drug addiction is clearly a secondary phenomenon.

It is possible, however, that the modern food environment of developed countries has created food addiction as another secondary phenomenon of age-old reward and motivation mechanisms in the brain. Michael Lowe and Meghan Butryn have defined something they call "hedonic hunger," in contrast to the "homeostatic hunger" that people and other animals feel when they are in need of sustenance.[61] Hedonic hunger has developed in response to "society-wide changes in both the physical and psychological availability of food [that] has created a type of eating motive in whole populations that has never been seen before."[62] This concept is similar to, if more general than, David Kessler's notion of conditioned hypereating.

Lowe and Butryn point out that the United States has had an environment of caloric abundance for decades, predating recent increases in obesity rates. They argue that recent changes in American society have encouraged a more permissive environment about eating, making it acceptable to eat at any and all times. This change decouples hunger from its homeostatic role and allows hedonic eating to become the primary form for many people. Like drug addiction and compulsive gambling, eating becomes only about pleasure, since if a person is never actually deprived of food, satiation is not really an issue. It seems to me that food addiction and hedonic hunger go hand in hand—both representing a shift in eating from being mostly about feeding the body to being mostly about feeding the mind.

Addicted to Not Eating

In Western developed nations, where 20–30 percent of the people are obese and many more than that overweight, and where the health consequences of too much fat are advertised and discussed on a regular basis, dieting is for many people an ongoing fact of life. People who successfully lose weight are congratulated, and commiseration is in order when someone fails to lose or gains back what he or she has lost. Losing weight and keeping it off are legitimately noteworthy achievements. But there are limits. A successful dieter must at some point reach a healthy weight that can be maintained by finding a balance between too much eating and too little.

In the psychiatric condition known as anorexia nervosa, the balance in maintaining body weight shifts, incongruously in environments of nutritional plenty, toward thinness.[63] The formal diagnosis of anorexia requires several features to be present: body weight at less than 85 percent of normal, an intense fear of gaining weight, significant disturbance in the perception of body shape and size, and (for women) amenorrhea. People with anorexia consume very few calories, sometimes limiting their diet to a small number of foods, and often use purging or excessive exercise to keep the weight down. The self-starvation of anorexia is compelling, especially in an environment in which warnings about the dangers of excessive consumption abound but are rarely heeded.

Weight loss can be regarded as heroic, but when an individual loses the ability to limit her weight loss (about 90 percent of anorexia sufferers are female) and goes beyond an acceptable threshold, she exceeds not only biological but social norms. Now, exceeding the biological norm of thinness can be deemed good or bad. The fasting girls of the Middle Ages were celebrated for their piety and devotion, and in contemporary society there are pressures for girls and women to be thinner rather than fatter.[64] But the extreme thinness of anorexia goes beyond ideals of attractiveness or piety, and so the behaviors associated with anorexia are regarded as psychopathological. Social factors aside, anorexia is unhealthy. Mortality rates for individuals with anorexia are much higher than those for age-matched controls, and anorexics are afflicted by a wide range of medical issues affecting the skin, bones, multiple internal organ systems, and metabolic and hormonal function.[65]

Anorexia is strongly associated with depression, and suicide rates among anorexics are very high. Anorexic individuals also exhibit high rates of autism-like disorders. These features are usually thought to be predispositions for the development of anorexia. However, anorexia itself has a marked effect on brain anatomy. Loss of gray matter in several parts of the cerebral cortex is significant in the acute phase of illness, and gray matter volume may not fully be restored even with recovery.[66] As we have already discussed, reduced gray matter is likely a contributor to the development of dementia in older age, so the effects of anorexia on brain health are potentially long-lasting. Cognitive deficits associated with anorexia can persist for years following apparent recovery from the disease.

When anorexia first came to public notice in Western countries in the 1970s and 1980s, there was a concern that it was a historically unique manifestation of an overheated media and advertising culture that put undue stress on adolescent girls and young women. However, historians such as Joan Jacobs Brumberg made it clear that the phenomenon of fasting girls was nothing new, and noted that in the twentieth century it was a combination of factors, including increases in female control and self-determination and the expanding dieting and exercise culture, that led to an apparently alarming increase in the rate of anorexia.[67] Compared to obesity, of course, anorexia is relatively rare. While the majority of people are prone to gaining weight, genetic studies indicate that genes have a significant role in the development and expression of anxorexia.[68] It is equally clear that environmental factors also play a role.

The notion of self-control is a central one in the cultural and environmental interpretation of anorexia. As Susan Bordo writes:

> The young anorectic, typically, experiences her life as well as her hungers as being out of control. She is a perfectionist and can never carry out the tasks she sets herself in a way that meets her own rigorous standards. She is torn by conflicting and contradictory expectations and demands, wanting to shine in all areas of student life, confused about where to place most of her energies, what to focus on, as she

develops into an adult. . . . Usually, the anorexic syndrome emerges, not as a conscious decision to get as thin as possible, but as the result of her having begun a diet fairly casually, often at the suggestion of a parent, having succeeded splendidly in taking off five or ten pounds, and then having gotten hooked on the intoxicating feeling of accomplishment and control.[69]

The phrase "gotten hooked on the intoxicating feeling" makes anorexia sound a lot like addiction. Studies of dopamine metabolism in anorexia do support the theory that there is something different in the reward system of anorexic brains.[70] Individuals with anorexia often exhibit hyperactivity. Although this is commonly thought to be related to the desire to expend calories to lose weight, Anton Scheurink and colleagues suggest that anorexic individuals become addicted to activity itself, perhaps in part as a replacement for the psychological reward more typically derived from the pursuit, preparation, and eating of food.[71] The dopamine circuits associated with motivation have long been shaped by evolution to encourage food-related activity.

Another neurotransmitter, serotonin, is also implicated in anorexia. Abnormal functional activity of serotonin may be linked to changes in mood and perception of satiety that are hallmarks of anorexia. Walter Kaye and colleagues hypothesize that "disturbances in the 5-HT [serotonin] system contribute to a vulnerability for restricted eating, behavioral inhibition, and a bias toward anxiety and error prediction, whereas disturbances in the DA [dopamine] system contribute to an altered response to a reward."[72] They suggest that these vulnerable systems, encompassing distinct brain pathways, are particularly susceptible to "dysregulation" due to the effects of female gonadal steroids and other age-related changes that accompany puberty. The cultural environment that condemns weight gain and celebrates weight loss provides a cultural stress, which in these particular at-risk girls and young women is managed abnormally, by restricting food intake and losing weight. Not eating is rewarding and mood-elevating. To paraphrase R. D. Laing, overeating is a sane reaction to the insane world of nutritional plenty and excessive consumption; anorexia is the insane response to that insane world.

But is the world in which anorexia is a notable public health problem one that is not just nutritionally flush but also Westernized? Studies of cultures in transition from traditional patterns to more modern and Westernized ones, of immigrant populations, and of minorities within Western societies all suggest that anorexia and other eating disorders become more common with greater exposure to mainstream Western influences.[73] For example, in the Pacific island nation of Fiji, extensive changes in body image perception and attitudes toward diet in girls were observed in the years immediately following the introduction of television in 1995.[74] Girls who watched television (with much Western programming) developed preferences for a thinner body type, which was accompanied by increased levels of body dissatisfaction, dieting, and behaviors such as purging.

Anorexia is substantially less common in developed, urbanized Asian societies compared to Western societies, but it is present. Increasing rates of anorexia in Japan after World War II could certainly be ascribed to Westernization, but researchers believe that factors within Japanese society also drive and shape the expression of anorexia. As in Western countries, thinness does figure into Japanese ideals of female beauty. However, the expression of anorexia in Japan is much less engaged with the goal of losing weight to be more attractive, or even with the reward of weight loss itself. For example, "fat phobia" and fear of weight gain are much less common in Asian women with anorexia compared to Western women and girls.[75] Instead, much more fundamental issues of self-control and self-determination come into play. Kathleen Pike and Amy Borovoy suggest that some Japanese anorexics have the goal of delaying maturation, and thereby delaying or avoiding entrance into the socially prescribed and, in the view of some, limiting and dependent role of "housewife."[76] Another case they report describes a Japanese woman who had gained weight while overseas but lost it upon her return to Japan, subsequently becoming anorexic. Her goal was not to be attractive but to simply fit in better with everyone else. What is clear is that wherever anorexia occurs, it is the result of the interaction between a vulnerable brain chemistry and the cultural and familial environment. Some environments are clearly more likely than others to push at-risk individuals over the edge into illness.

One of the bedrocks of human sociality is eating. We do not just share a meal or a feast: we hunt and gather food in groups, we prepare it with family and friends, and when we are not eating or preparing food with others, we spend quite a bit of time talking with them about food and the next meal. In anorexia, the lonely and misguided pursuit of damaged perfection, rejecting food also means rejecting a critical aspect of human social life.

Anorexia is not a global problem in the way that obesity is in developed countries or chronic hunger and malnutrition are in many developing countries. But it illustrates the mind-food connection in a particularly poignant way. As individuals, we are likely to perceive our own relationships with food as being intuitive, like the competence we have with our first language. It is clear, however, that our mental model of how to eat is, like language, an instinctual process (subject to individual genetic variation) that requires and is shaped by a nurturing environment. Anorexia represents an extreme version of how to think and eat food, but it is on a continuum with the myriad of mindful ways that all people use when they come to eat.

Mind and Dieting

This chapter began with an existential lament about fat and a nineteenth-century invitation to consider the many facets of human folly as they relate to the phenomenon of extremely thin girls. The trials and tribulations of weight gain and loss are familiar to many who live in the urbanized developed world. Some foods are considered "good," others "bad." Some diets are healthy, while others are guaranteed to send you to an early grave. The ancient Etruscans and Romans used to read the entrails of sacrificed animals in order to foretell the future. Today, it's the numerology of dieting that holds sway: the numbers involved in weight-cholesterol, triglycerides, fasting glucose level, and so on are consulted so that we may live longer and healthier lives. There is nothing inherently wrong with this as long as it is recognized that at an individual level, the predictive ability of these numbers is either intrinsically limited or swamped by the many other determinative and random factors that contribute to a person's life span.

Paying so much attention to these diet-related numbers can be quite stressful. Since stress itself is not exactly conducive to living a long and healthy life, can we conclude that people would be better off not paying attention to them? Perhaps, but there is no denying that obesity increases the likelihood of developing many illnesses, and being ill is itself quite stressful. If a person is too fat, assessed by whatever reasonable and conventional measure one wants to use, then he or she probably would be better off losing some weight.

Illness is not a moral failing, and to the extent that it contributes to illness, obesity should not be considered a moral issue. We have looked at how eating is embedded in the evolution of our sociality and hence forms the foundation of nearly all human cultures. If food is traditionally so much about family and friends, sharing in both plenty and scarcity, then it is easy to see how mis-eating can be construed as misbehavior on a moral level. We no longer live in that traditional cultural world, however, and eating too much should not be seen as shameful or immoral. That is a misplaced stress, derived from the evolved social psychology of eating, and it does not make losing weight any easier. We should also recognize the extent to which our evolutionary cognitive psychology rewards us for eating more rather than less of those foods that we know make it so easy to gain weight.

If we could go back in time and survey all of our ancestors over the past 6 million years, asking them, "What will be the biggest problem our species might face in the future?" it is highly unlikely that very many of them would respond with "Too much food." It is truly extraordinary that billions of people today live in an environment of ongoing caloric plenty (while, of course, billions of others do not), and hundreds of millions of them are apparently suffering for it. There is a mismatch between the human body and the present-day food environment, and that mismatch is mediated by the mind. Two things are certain: first, barring a collapse of the global food economy, almost everyone in developed and newly developing countries will continue to have access to abundant cheap calories; and second, human bodies are not going to evolve quickly enough to be able to physiologically handle these excess calories in a salubrious way. That leaves the mind, shaped by natural selection to be flexible and adaptable, as the best target if we want to address the obesity epidemic. Dieting involves changing not just what is eaten but also how food is thought and mentally processed. The better we understand the cognition of food and eating, the easier it will be to effect these changes in diet that so many public health officials say are desperately needed.

5

MEMORIES OF FOOD AND EATING

When it comes to food, I don't have a gourmet's memory.
I remember the kinds of food I was raised to love. Chaz and
I stayed once at Les Prés d'Eugénie, the inn of the famous Michel
Guérard in Eugénie les-Bains. We had certainly the best meal
I have ever been served. I remember that, the room, the people at
the other tables . . . but I can no longer remember what I ate. It
isn't hard-wired into my memory.

   Yet I could if I wanted to right now close my eyes and
re-experience an entire meal at Steak 'n Shake, bite by bite in
proper sequence, because I always ordered the same items and
ate them according to the same ritual. It is there for me.

—ROGER EBERT, *Life Itself* (Grand Central Publishing, 2011)

WHENEVER I EAT THE FRENCH CAKE-LIKE COOKIE known as a
madeleine, I am immediately transported back to a freshman
comparative literature class I took at Berkeley. We gathered that
warm spring day on a lawn bordering Strawberry Creek, and the
instructor brought us madeleines to help us understand how they
served as a mnemonic launching pad in Proust's *Swann's Way*.
Beyond the madeleines, I don't remember all that much about
Swann's Way. I don't remember the instructor's name, but I do
recall that there was a young woman in the class, a friend of a
friend, whose boyfriend was a water polo goalie; he had the widest

thumbs I have ever seen. Alas, it is about at this point that my madeleine-inspired trips down memory lane typically end, interrupted by a pointed question from my wife: "How many of those have you eaten?"

We all have our food memories, some good and some bad. The taste, smell, and texture of food can be extraordinarily evocative, bringing back memories not just of eating the food itself but also of the place and setting in which the food was consumed. Beyond memories of taste and place, food is effective as a trigger of even deeper memories of feelings and emotions, internal states of the mind and body. These kinds of memories can sneak up on you: they have the power to derail a current train of thought and replace it with one both unexpected and unexpectedly potent.

Not all memories are created equal, and this is true whether or not food is involved. The passage from film critic Roger Ebert at the beginning of this chapter contrasts the nature of his memories of a critically exalted dining experience with those from a more familiar but mundane meal. Because of cancer surgery Ebert can no longer eat (or talk), and thus his memories of food can no longer be evoked by eating or the prospect of eating. But as he recovered from surgery and adjusted to his changed life, he found that some food memories were particularly poignant for him and others were not. The "event" meals he enjoyed over the course of his life were indeed memorable, but the foods were not. Instead, the foods he enjoyed often while growing up, such as fast-food burgers, candies, and sodas, are the ones that he can imagine eating with surprising clarity. Repeated exposure to these foods probably has something to do with why his memories of them are so vivid. But his memories of them may also be so powerful because he formed them when he was younger, at the time when his whole cognitive outlook on food was developing.

A person who plans to eat again does not have to engage in such mental gymnastics, but it's an exercise I think everyone should try. You will likely find that the powers of recollection are surprisingly acute, and that the multisensory experience of eating lends itself to the remembrance of details both large and small. (To hark back to the previous chapter, one of the details of eating that Ebert misses most involves not food but the loss of social interaction that goes along with it.)

Cognitive scientists recognize many different types of memory: short-term, long-term, declarative, episodic, explicit, implicit, retrograde, anterograde, procedural, working, prospective, and so on. At a higher level, we have memories that are collective, cultural, and traditional. Memories can be repressed and recovered. The phenomenon of memory slides into other cognitive categories, such as learning, intelligence, and conditioning. Memory is essential to creating and maintaining the autobiographical narrative that defines the self.[1]

Some of these types of memory have a physiological basis, and some are categories of convenience for behavioral scientists. The heightened sense of memory we often experience in connection with food raises a couple of questions: Is food in any way a privileged part of the environment and personal experience when it comes to memory? Are our brains primed to form food memories more easily than other kinds? For all animals, finding food is one of the critical aspects of survival, and memory ability can certainly be shaped by natural selection in relation to food and eating. On the other hand, we expect that memory systems will be critical to all sorts of things an animal might want to do, so food acquisition would be one selective force among many. For humans, whose culture and consciousness provide the collective power to isolate and promote certain aspects of the struggle for life over others, food (along with sex and status) has become one of the touchstones of social life. Food memories are important not just because they concern sustenance but also because they have extensive connections to other memories of people, places, and things.

Hippocampus Means "Seahorse"

The hippocampus is a brain structure (or two brain structures, one in each hemisphere) that is critical for the formation of declarative, or explicit, memories.[2] These are memories that we can consciously recall—the ones we can talk about and remember as facts or events. In humans, declarative memory is tied up with language: we literally give voice to these memories, either out loud or in our own minds. Declarative memories are not unique to humans; we just have the ability to declare them. In other mammals, hippocampal-dependent memories are "characterized by rapid formation, complex associative properties, and flexible expression."[3] In humans, the complex associations and flexible expressions of these kinds of memories, combined with our well-developed executive functions, contribute to our unprecedented ability to turn previous experiences into future plans and actions.

The hippocampus itself is a complex structure, which was given its name because of its (faint) resemblance to a seahorse (the genus name for all seahorses is *Hippocampus*, derived from the Greek words for "horse" and "sea monster"). It is composed of a curved and folded sheet of cortex that is tucked into a gyrus (ridge) of the temporal lobe along the interior surface of the hemisphere; a distinct head and tail are its most seahorse-like attributes. Within the temporal lobe, the hippocampus lies with its head oriented toward the front of the brain, the tail running behind it.

The hippocampus is part of the limbic system, the complex of brain structures responsible for the regulation of drives and emotion, among other things.[4] The limbic system also includes the hypothalamus, which we discussed earlier, as well as the amygdala, a cluster of neurons that sits just in front of the hippocampus within the temporal lobe. The amygdala is concerned with emotional regulation and control. Elsewhere in the cortex, the anterior cingulate cortex is also part of the limbic system and, as we discussed earlier, likely has an important role in incorporating emotion into higher processes such as decision making. In addition to its role in memory, the hippocampus is important for spatial orientation and navigation.

Via the surrounding cortex of the temporal lobe, the hippocampus has inputs from all senses. Some of the primary cortex for the sense of smell, or olfaction, is located in the temporal lobe

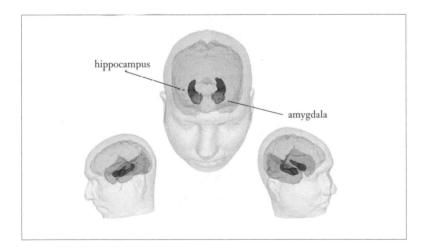

Two components of the limbic system located within the medial temporal lobe: the amygdala, important for emotional regulation, and the hippocampus, essential for memory, among other functions.

near the hippocampus, so there are particularly intensive connections for this sense. The hippocampus also receives fairly direct inputs from the amygdala. The limbic system was once considered to be a kind of "primitive" brain buried within the more advanced cerebral cortex; its involvement with olfaction, basic drives, and emotion was seen to reflect its control of these more basal, primitive functions. We now know that brain networks integrate different parts of the brain, old and new, primitive and derived, and thus it is misleading to think of more advanced brain structures being layered over more primitive ones.[5] Yet the close functional links between the hippocampus and primary emotional and olfactory regions may help explain why some emotional memories are particularly strong and why some odors trigger memories very quickly.

The role of the hippocampus in forming new memories has been most strikingly revealed in studies of patients who have had their hippocampi destroyed by surgery or illness.[6] The long-term memories already stored by these individuals are readily available to them, usually up to some fairly recent time preceding the loss of the hippocampi. They remember the past clearly and have just as strong a sense of their own autobiographical self as anyone. However, they can no longer form new declarative memories. For the rest of their lives, their "present" is locked at the time when they stopped forming new memories; no new experiences can become a conscious part of their store of memories. As they age, a look in the mirror can be distressing, as the image in the reflection does not match the youthful person they remember. Like everything else, however, the distress is soon forgotten.

Individuals who have lost only part of the hippocampus because of disease or injury can often retain some of their ability to form new declarative or explicit memories, and it appears that the preservation of declarative memory can be related to the amount of hippocampus that remains.[7] But an intact or partially intact hippocampus is not necessary to form some kinds of memories. Procedural memories are a kind of implicit memory; the term *procedural* refers to the ability to learn and retain the motor skills necessary to do a specific task or action. Injuries or lesions to the motor areas of the brain, such as the basal ganglia and the cerebellum, disrupt the ability to form new procedural memories. However, even individuals with the most profound amnesia from loss of the hippocampus can learn new motor tasks, even something as complex as using a weaving loom, suggesting that their procedural memory system is fully intact.[8] Of course, the amnesics cannot remember the repeated sessions in which they were instructed and practiced the tasks; they sometimes express surprise that they were able to do something so complex on their first try.

The diverse origins of declarative and procedural memories suggest that there is no single brain region where memories are formed, nor are there specific neurons that match up one to one with each memory. Instead, memories reside in networks distributed throughout the cortex, where they are organized hierarchically according to their complexity and abstraction. Based on decades of research, Joaquin Fuster has proposed a comprehensive network model of memory storage, which has several basic principles: memories are stored in widespread cortical networks, formed by experience; more complex memories, such as an autobiographical memory, necessarily link cells from different brain regions; memory networks overlap and link with one another via common nodes.[9] Neurons form associations leading to networks when they are activated simultaneously; networks are also formed when a single neuron receives simultaneous inputs from more than one other neuron. The individuality of memory is a result of the fact that there are billions of neurons in the cortex, whose network connections are subject to the influences of numerous variables. Memories do not reside in single cells, but there are "memory cells" found in many cortical regions that form parts of networks.

This gives us a bit of an idea about the nature of memory in the human brain. I will discuss some other forms of memory as we go along, but let's consider Roger Ebert's recollections again in light of these brain memory mechanisms. First, he says that he does not have a "gourmet's memory"—presumably he has a film critic's memory. But what this means is that when it comes to food, he does not have the food expert's ability to easily note and retain new food-related information. If we accept the network model of memory storage, then an expert who deals repeatedly with a certain topic will establish and reinforce memory networks associated with that topic. New memories are efficiently created because they slot into well-established networks and associations. Such an ability can appear effortless to outsiders, yet in reality it is a reflection of years of engagement with a topic.

Ebert describes his dinner at Les Prés d'Eugénie in glowing terms: he is with his wife, it is the "best meal" he has ever had, he remembers the setting with great fondness. The vividness of the memory is no doubt a result of its emotional content; the strong connections between the hippocampus and the amygdala tend to potentiate the creation and strength of emotional memories. Even if the meal is not remembered in detail, having a "best meal" would be an emotional experience for anyone who likes food. Combine an extraordinary series of dishes with an extraordinary setting and an extraordinary person, and you have the recipe for a memorable experience. It may seem odd that the food—the centerpiece of the experience—has been forgotten, but clearly the food was less important in the context of this particular emotion-laden event. In a different setting, or in a person with a different set of memory networks, emotion might aid in better remembering the food, good or bad.

Finally, consider Ebert's mental re-creation of eating a Steak 'n Shake burger. I suspect that Ebert, like many American males, is an expert in eating not just Steak 'n Shake burgers but hamburgers in general. He could probably from memory (unless he was monomaniacally loyal to Steak 'n Shake) compare and contrast their burgers with those of many other chains. It is with some distress (and a little pride) that I realize that I can, off the top of my head, remember in detail burgers from McDonald's, Burger King, Wendy's, White Castle, Hardee's, Carl Jr.'s, Jack in the Box, Five Guys, In-N-Out, Hamburger Habit, Nation's Giant, and the now defunct Rich's Bulky Burgers. Interestingly, I have quite a strong memory of my one and only visit to a Steak 'n Shake: I was with my older son, it was after a morning session of a swim meet in the Cincinnati area, it was Saturday, and I was surprised by the fact that they had sit-down service. We were both excited to try the burgers for the first time, and alas, I remember that we were somewhat disappointed (perhaps our expectations were too high). Like Ebert at Les Prés d'Eugénie, I remember the quality of the food, the company, and the setting, though a clear memory of the burger escapes me. But to get back to Ebert's ability to re-create from memory the experience of eating a Steak 'n Shake burger: this is in effect both a declarative memory and a procedural one, of a happy event, involving multiple sensory modalities, reinforced through countless iterations. No wonder he can summon the experience at will.[10]

Declarative memories about food are just that; they are not in some way physiologically unique "food memories." Antonio

Damasio writes that autobiographical memories "are likely to use the same sort of framework used for the memories we form about any entity or event. What distinguishes those memories is that they refer to established, invariant facts of our personal histories."[11] Thus many of our food memories can be classified as autobiographical memories concerning food. This does not mean that food may not be a powerful evocative force for memory creation or recall. The strong connections between the hippocampus and the olfactory and emotional centers of the brain may predispose us to an essential link between food and memory. Indeed, such a connection could be very adaptive for any mammal.

The Hippocampus, Eating, and the Amuse-Bouche

Larderhoard and *scatterhoard* are two truly wonderful words. They refer to two ways that animals store food (in collections referred to as caches) for recovery and eating later. Many species, such as squirrels, are wholly dependent on cached food for survival over much of the year. Larderhoarders store large amounts of food in a few locations, while scatterhoarders place smaller amounts in many locations. Even within a species, populations can vary, depending on ecological conditions, as to whether they are more scatter- or larderhoarders.[12]

Food hoarding is of particular interest to scientists interested in the hippocampus because it offers a way to test ideas about the relationships among food, memory, and the anatomy of the hippocampus. Hoarding, or caching, depends on two primary hippocampal functions, memory and spatial navigation. One way to examine the relationship of memory to eating behavior is to compare the size or patterns of cell growth of the hippocampus across species, or even within species. Do these aspects of hippocampus anatomy reflect the memory demands of their diets? Hippocampus shape actually varies quite a bit between species—the seahorse shape of ours is not typical—but comparing the hippocampus among different species is legitimate because at the cellular and functional levels there is much similarity.

Several kinds of animals have been studied to tease out the relationship between hippocampus anatomy and feeding behavior.[13] Birds such as black-capped chickadees, nuthatches, and jays store food in numerous locations throughout their range. Compared to bird species that do not store food, these three species have larger hippocampal complexes relative to both overall brain volume and body size, confirming in this instance that increased volume goes along with increased usage and ability. A similar pattern can be seen in rodents, where kangaroo rat species who store food in a single burrow have smaller hippocampi compared to similar species who store food in dispersed locations. And among North American red squirrels, eastern scatterhoarding populations have increased size in part of the hippocampus compared to those from western larderhoarding groups.

In humans, no such variation has been found in hippocampal size that is related to diet or eating habits (not that anyone has looked too hard). The absolute size of the human hippocampus is about what would be expected for a primate our size; in comparison with other animals, however, the human hippocampus takes up a relatively small portion of the brain, as our overall brain volume is much greater than would be predicted for our body size.[14] Interestingly, there is some evidence that the size of the hippocampus is inversely correlated with performance on IQ tests—in this case, less is more.[15] Presumably, neural "pruning" increases the efficiency of the hippocampus during brain development or with experience, perhaps by removing redundant connections between neurons.

Studies in people from various occupations indicate that the size of the hippocampus is quite plastic, especially in response to spatial navigation training. The famous studies of London taxi drivers, who must master the complex streetscape of London before being licensed, demonstrate that they show an increase in size of the posterior part of the hippocampus compared to ordinary drivers, but the anterior part is smaller.[16] This pattern becomes more pronounced as the taxi drivers become more experienced. Although learning London's streets is a heavy-duty memory task, it appear that the spatial expertise involved in the task is what drives the changes in the hippocampus. This is supported by the fact that physicians, who also must memorize tremendous amounts of material, show no change in the anatomy of the hippocampus with more or less experience or compared to people with matched IQs who had no tertiary education.[17]

Diet-related demands on spatial memory can clearly influence variation in the size of the hippocampus in various species. But remembering where to find food is important for all animals. How does the body link ingestion and memory via the hippocampus? It comes as no surprise that the body's chemical messengers, the hormones, are likely involved. The hippocampus is rich in receptors for several hormones, including insulin, leptin, and ghrelin, that are active in both the gut and the brain. This suggests that the hippocampus is directly influenced by the forces of appetite, or may even assist the hypothalamus in directly regulating appetite.[18]

Ghrelin is a hormone that functions in some ways in the opposite direction to leptin. At high circulating levels, leptin suppresses appetite and eating is reduced; when ghrelin is given to laboratory animals, they eat voraciously. Both hormones work toward the same goal: to signal when energy stores are low and an animal should begin feeding. For example, people with anorexia have low levels of leptin, signaling low energy stores, and high levels of ghrelin, which should increase their appetite and encourage feeding.[19] Clearly, part of the problem in anorexia is that other psychological factors override the normal response to high ghrelin levels.

Animal studies indicate that leptin and ghrelin both appear to have direct effects on the anatomy and function of the hippocampus and on memory performance.[20] Leptin promotes the formation of synapses in the hippocampus and other memory-enhancing processes. Ghrelin has similar effects, which are enhanced with increasing levels of the hormone. Because seeking food is so critical to animals' survival, it comes as no surprise that these behaviors should be hormonally linked to memory.[21] Insulin also has a positive effect on memory and hippocampal function. The increased risk of dementia in older obese individuals may in part be due to the development of insulin and leptin resistance (which are also responsible for type 2 diabetes). Although under normal conditions higher levels of these hormones may enhance memory, when the body no longer responds to these hormones, even very high levels can have no positive effect.[22]

There appears to be little doubt that hormones that regulate eating and appetite in general can affect the hippocampus and memory function. Given this, we might wonder whether there is anything we can eat that will enhance our memory abilities. One component of many foods, glucose, has been repeatedly shown to be a memory enhancer in lab studies, for both animals and humans.[23] To some extent, this may be a result of the fact that glucose is the main fuel of the brain and therefore should be expected to improve its performance in all domains; as attention and arousal increase, memory may be improved. During some memory-dependent cognitive tasks, the hippocampus may be intensely active, and therefore memory performance would benefit from an increased supply of glucose. These are all more secondary rather than primary improvements to memory, however. Spiking insulin levels resulting from a glucose load could also account for some of the memory-enhancing effects of glucose. Another sugar, fructose, can enhance memory performance, too. But unlike glucose, fructose does not cross the blood-brain barrier, so there may be a yet-to-be-determined peripheral mechanism for memory enhancement that is shared by both glucose and fructose.[24]

Another food component that may improve memory is the flavonoids. Flavonoids are secondary compounds produced by some plants; they do not play a direct role in the plant's growth or reproduction, but they are useful for defense against herbivores. Flavonoids are found in tea, cocoa, and fruit such as citrus, grapes, and blueberries.[25] More than 6,000 flavonoids have been identified, and animal studies indicate that they may slow age-related memory loss and improve other aspects of cognitive performance. Although how flavonoids might improve memory is not well understood, it is likely that the combined effects of enhancing synapse formation, improving cardiovascular health, and protecting neurons via antioxidant and anti-inflammatory activity all contribute to enhancing memory and cognitive performance. Studies of flavonoid therapy in humans have produced some intriguing evidence of memory improvement, although it is still too early for a definitive picture to emerge. It is probably worth exploring the idea some more, because if such a readily accessible dietary component can improve cognitive health in older individuals, the economic impact could be tremendous, especially in a population where elderly individuals make up an increasing percentage of all adults.[26]

Another commonly available non-nutritional dietary substance, caffeine, may also have profound effects on memory.[27] Caffeine is, of course, a drug, widely consumed in coffee, tea, and soft drinks. Epidemiological studies indicate that coffee consumption may have a protective effect against developing Alzheimer's disease. Lab studies on mice support the hypothesis that caffeine can enhance memory, and furthermore suggest that caffeine may both protect against the development of Alzheimer's and reverse some of its effects, at least in mice. The mechanisms whereby caffeine may improve memory are probably generally similar to those that may operate with the flavonoids; in addition, caffeine may specifically interfere with some of the physiological processes that lead to Alzheimer's. Another possibility is that caffeine may promote the production of cerebrospinal fluid, leading to increased activity of enzymes important for cellular energy production and stimulation of cerebral blood flow.[28] Evidence for the therapeutic efficacy of caffeine in Alzheimer's disease is so strong that there should be more clinical trials on caffeine as an inexpensive treatment for the disease.[29]

At this point, your memory is probably being taxed by all this discussion of memory mechanisms. To recap the anatomical and molecular underpinnings of memory, we see that many factors can influence the creation and strength of food memories. The emotional context plays a role, as do the direct sensory impacts afforded by taste and especially smell. How hungry you are when you eat also plays a role, as the hormones leptin and ghrelin signal the brain, including the hippocampus, to maintain proper energy stores and encourage eating. The effects of glucose and insulin on memory suggest that what you eat may influence your ability to remember it. Flavonoids and caffeine may have long-term effects on the health of memory, which of course may influence people's ability to remember food events in their lives.

To finish up, let's try to apply some of this knowledge about food and memory toward explaining the popularity of a relatively new innovation in meal service, the amuse-bouche. In French, *amuse-bouche* literally means "mouth amuser." The amuse-bouche usually consists of one or more bite-sized treats, served gratis at the start of a meal. It is not ordered by the customer but presented as a kind of gift from the chef, who chooses what it will be. Although small, the amuse-bouche offers an opportunity for the chef to highlight his or her technique and creativity right at the beginning of a dining experience; in many cases, these little foods are as elaborately prepared as any other course during the meal.

The amuse-bouche is a product of the nouvelle cuisine movement of the 1970s, and by the 1990s it was becoming standard at high-end restaurants; now it has moved down the restaurant food chain to the less fancy places.[30] Since restaurants are in the business of selling food, the popularity of the amuse-bouche can be taken as a sign that these small treats are helping to achieve this goal. No doubt one reason it works is that everyone likes a gift, a little something for nothing. A more general attribute of the amuse-bouche, I think, is that it makes a given restaurant experience more memorable.

Consider that when a patron enters a restaurant, especially one that is known for the quality of its food, he or she is in a heightened state of expectation, and hungry as well. The amuse-bouche builds on this emotional foundation by providing a surprise, and it also can be seen as a homely or familial gesture from the kitchen. All of these elements help set the cognitive table for a memorable meal. Being hungry means that one's circulating ghrelin levels are higher, enhancing memory formation.

From a memory standpoint, the first few minutes of a restaurant meal are critical, before hunger is sated and emotions are calmed. The amuse-bouche can carry the distinctive stamp of a restaurant when the patron is primed to remember it. An amuse-bouche that includes a hint of sweetness might make the meal even more memorable, depending on how quickly a patron's insulin kicks in. Note that I am not saying that the amuse-bouche itself is memorable; rather, it lays the groundwork for remembering the whole restaurant experience. However, it should make an impression at the time; otherwise, like the bread or rolls that traditionally arc offered when a patron is first seated, it is just calories that deaden hunger and diffuse the emotion of what should be an exciting experience.

Eating, Remembering, and Forgetting

Imagine that you are a typical social primate, a monkey or an ape. What are some of the qualities of food sources and eating that you might want to remember? Over the long term, it would be useful to remember things such as location, the quality of the food, seasonal availability, the likelihood of encountering predators at the site, and the chances of encountering other animals trying to eat the same food. None of these would be in the form of conscious, declarative memories, but instead would be stored as knowledge accrued with experience. Over the course of evolution, it is understandable that natural selection has shaped memory in relation to food seeking. However, food *eating* would be another matter. Once the animal is at a food source, the goal should be to eat as much as possible before the food source runs out, or daylight runs out, or something comes along to scare the animal away. If none of those things happen, then the animal should eat until full but not immobile. But memory of the quantity of food eaten would not be a particularly useful ability other than in the sense of all, part, or none.

Humans are the only primates who might find themselves in a position where they are forced to recall what and how much they have eaten. How good or bad people are at doing this has been of great concern to epidemiologists who want to draw connections between diet and health or disease. Research in this area, the results of which are often widely reported in the popular media, depends on the use of dietary surveys, food frequency questionnaires, and other instruments that require people to recall their patterns of consumption. These various instruments can be tested under highly controlled situations in which the amount of food consumed is measured or physiological markers of dietary intake are assessed.[31] However, it is impossible to do such careful studies with large groups of people over long periods of time. The large numbers of individuals that must be surveyed in order to find correlations between diet and disease means that epidemiologists have little choice but to depend on individual recall to get dietary information.

This state of affairs has led to some controversy in the field, with researchers being more or less dubious about food frequency data and the conclusions that can be drawn from them.[32] David Paul and his colleagues conducted a very careful study in which the diets of a dozen men were carefully monitored over a sixteen-week period. At the conclusion of the monitoring period, the subjects were given a standard food frequency questionnaire and asked to recall how they had eaten (they were given detailed instruction on how to fill in the form, which was actually meant to cover a year's recall, not just sixteen weeks). The upshot of all this is that if we might expect a group of people to be able to accurately recall what they ate over a moderate period of time, then this was the group: they were all healthy and reasonably well educated, they knew they were in a nutritional study, they knew that their food intake was being carefully monitored, and their food survey responses were vetted for obvious errors. The results, however, simply reaffirmed that even under the best circumstances, people have poor memories about what they eat. Some of the dismal observations made by Paul and his colleagues: "Absolute and relative macronutrient intakes were poorly predicted by the food frequency questionnaire. . . . Given our homogeneous study population . . . there was surprisingly large subject-to-subject variation in measurement error. . . . The usefulness of food frequency questionnaires to quantitatively measure the relation between disease and food intake in any size study is questionable."[33]

To be fair, although some of these conclusions are quite negative, Paul and his colleagues also pointed out that with various corrections for energy expenditure and body weight, the data from food frequency questionnaires can be improved. Everyone acknowledges that these kinds of data are not perfect, but as Walter Willett, one of the most prominent diet and disease scientists, says, since "large-scale prospective studies are highly desirable . . . self-administered questionnaires are usually a practical necessity."[34] Where it is possible to use direct physiological measures or other corrections (and this is not necessarily doable), that is all to the better.

So we are left with the fact that people's ability to recall what they ate over the medium or long term is at best mediocre, and thus is bound to compromise scientists' ability to draw firm conclusions about the relationship between diet and disease. Are we any better at short-term recall? Psychologist Brian Wansink's studies suggest that we are as hopeless in that time frame as in a longer one.[35] He found that five minutes after leaving an Italian restaurant, 31 percent of patrons could not remember how much bread they ate and 12 percent denied eating bread even though they had. In another study, Wansink and his colleagues set up a Super Bowl party and invited hungry MBA students to it. They were given free access to a buffet featuring buffalo wings, and they could have as many as they liked over the course of the game, free of charge. Bowls to collect the discarded bones of the wings were provided to all tables. However, the bowls were regularly cleared from only half the tables, while for the other half, the bones were allowed to accumulate. With no visible evidence of their past consumption in front of them, over the course of the game the students who had had the bones cleared ate 28 percent more wings than those who were reminded of their past consumption. As Wansink concludes, "Our stomach can't count and we don't remember."[36]

The remembering part of the quote is a bit of an overstatement. The students at the party presumably would remember, at least for a while, that they had eaten chicken wings, even if they would have only an imperfect idea of how many. But what happens when people who cannot remember anything eat? Do we expect some sort of mechanism, other than memory, to kick in to remind them that they have eaten?

Eating studies of densely amnesic individuals make it clear that memory, or the lack of it, influences eating patterns. These are people who have damage to the hippocampus and other brain regions resulting in a profound loss in their declarative memory capacity. Although there is of course individual variation in appetite independent of memory, the amnesic individuals converge on a common pattern in these eating studies.[37] The experimental setup goes like this: When offered a meal or the opportunity to construct a meal from a selection of items, amnesics eat pretty much as anyone else would. Then, following completion of the first meal, all evidence of it is removed; after fifteen minutes, the subjects are again offered the identical meal. The amnesics do not recall eating the first meal, and typically have no trouble finishing this second meal. In general, there is little fall-off in the number of calories consumed in going from the first to the second meal. A third meal is offered, and in some cases, amnesic subjects will make some effort at eating it, although they usually are getting a bit full by this time. The amnesic individuals can also show the effects of satiety: having eaten a quantity of a food, they will report that it is less pleasant, even though they do not remember having just eaten it. However, this does not necessarily prevent them from eating another meal. Hyperphagia in amnesia is not just a laboratory phenomenon—an amnesic subject once "overdosed" on bananas at home and got sick, as a family member reported.[30]

What is going on here? Clearly, memory has some role in informing us about when we should start and stop eating. One thing to keep in mind is that circumstances other than hunger and satiety affect how much we eat. Most meals do not end when we absolutely cannot eat another bite. There are many social factors that define the setting of a meal, including cues as to when it starts and finishes. An amnesic subject comes into a room and is served a meal; he or she does not remember having eaten recently, thus goes along. By the time a third meal is offered, the feeling of fullness may be present, but without memory of eating, and in the context of someone serving a meal, it is understandable that eating is attempted, especially if the food is considered to be palatable. The internal cues we use to assess our eating and appetite have all developed in the context of possessing an explicit, declarative memory.[39] Without it, these feelings and sensations remain, but they are unmoored as the amnesic drifts in a continuous present.

There is one kind of meal or food that most people can remember for a very long time—the kind that makes the eater nauseous and perhaps vomit either soon after or within a few hours of consumption.[40] A classical example of conditioned learning, strong food aversions can develop after only a single exposure to a sickness-inducing food (I personally am wary of diner food in western Ontario following one unfortunate meal I had on a road trip long ago). This kind of food aversion is not the typical way of developing a dislike for a food. The vast majority of foods that people dislike did not at some point make them sick, and we can come to like foods that we initially did not like.[41] But it is much harder for people to come around to foods that initially made them sick, or which they associate with a bout of nausea. The food itself does not necessarily have to be the culprit. I know someone who ate a package of Oreo cookies while on a boat trip, then became seasick and threw up; Oreo cookies have no attraction for him now.

Food aversion in humans probably results from some combination of explicit and implicit memories. Studies on lab rodents make clear that taste aversions can be induced without a working hippocampus, so clearly there is an implicit component at work.[42] Indeed, many people may have implicit aversions to certain foods. These aversions could develop without their conscious awareness, and so they have no declarative memory about the events that caused them. I am not aware of food aversion studies on humans with hippocampal damage; such studies would be ethically questionable, of course.

The brain mechanisms whereby an aversive taste experience becomes a memory are not yet well understood. The insula, that island of cortex buried within the frontal lobe, may have some role to play.[43] The insula includes part of the gustatory cortex, and it is the site for the integration of visceral and taste information, leading to the formation of long-term memories about both aversive and pleasant taste events. It is likely that pleasant and unpleasant tastes have somewhat different pathways. The aversive pathway involves part of the amygdala, the emotional center of the brain. Given the strong association between emotion and the formation of declarative memories, this connection may be one reason why aversive food memories are so potent.

The nausea and vomiting induced by "bad" food is an important defense mechanism for animals against toxins in food and the environment. To some extent, human culture and learning render this mechanism less important for us, at least as a typical means of developing food preferences and avoidances. However, over the past million years, humans and our ancestors, cultural animals all, have shown a willingness to enter and explore new environments. These environments contained unfamiliar plant and animal foods that would have to be tested before being introduced to the larder. Basically, these migrants had one way to test the safety of a food: someone would have to taste it and see if he or she got sick. The great advantage of human culture and language is that it was then possible to share the results of such a test with friends and family.

Working Food with the Working Memory

Chimpanzees use a wide range of tools in order to obtain and process food.[44] Jane Goodall's seminal discovery of how chimpanzees collect and eat termites provided the first hint that humans were not the only primates who could think about, make, and use very specific tools to do a very specific task (i.e., "fishing" for termites by inserting a properly modified stick into a hole to reach deep into a termite nest). Chimpanzees also initially use a large stick to gain better access to insect nests, after which they insert a smaller fishing stick into the nest; thus they can use a combination of tools to accomplish a single task. Some chimpanzees use hammerstones to open up hard nuts, while others break down large fruit by hammering a wood or stone "cleaver" against fruit sitting on a stone "anvil." They spontaneously employ all sorts of items in their environment to accomplish tasks involving reaching, poking, and batting.

When a chimpanzee considers a food task, identifies a food source, selects a tool, uses the tool to obtain the food, and eats the food, he or she is relying on the capacity for what is now generally referred to as working memory. A simple definition of working memory is "the ability to maintain and manipulate information over short periods of time." Psychologist Alan Baddeley's very influential model of working memory breaks it down into several cognitive components.[45] Overseeing it all is something he refers to as the "central executive." This includes a variety of cognitive processes that are necessary to accomplish a goal or task, which in crude but not totally inaccurate terms could be seen as the little person in your head who runs things. Below the central executive are at least three distinct subsystems. The "visuospatial sketchpad" is a temporary storage area where visual and spatial sensory information is made accessible to the central executive. The "phonological loop" is where sound information is brought into working memory. In humans, the aural component of working memory is dominated by language, language processing, and the ability to store and use linguistic information to accomplish short-term goals. Linguistically "tagging" items in the environment or concepts or feelings helps us remember them. Since other animals do not have language, the phonological loop provides the basis for a qualitative difference in working memory between ourselves and even close relatives, such as the chimpanzee. A final subsystem, the "episodic buffer," is where information from the other two subsystems is integrated along with information stored in long-term memory (both declarative and procedural) and made available to the central executive.

Baddeley's concepts have been extraordinarily productive over the past four decades, inspiring much work into the components of working memory. Building on Baddeley's work, Thomas Wynn and Frederick Coolidge argue that enhancements in working memory form the basis for the evolution of the "modern mind."[46] The neural networks involved in working memory are by their very nature diffuse and widespread, so it is difficult to point out particular evolutionary developments in the human brain as being critical for their evolution.[47] Of course, this is true of any of the complex cognitive adaptations expressed by the human brain. An exception to this is language, where we can at least localize the parts of the brain specialized for the motor control of speech, speech comprehension, and other aspects of language. These dedicated language networks may provide insights into how working memory evolved differently in our brains compared to similar processes in other animals.

Wynn and Coolidge argue that the archaeological record suggests that an enhanced working memory developed relatively late in human evolution (on the order of tens of thousands of years ago, in fully modern *Homo sapiens*). Other researchers, such as Philip Beaman and Miriam Haidle, have compared chimpanzee termiting with the use of the simplest stone tools from 2 million years ago to cut meat.[48] Using different, formal, step-by-step analyses to model these tasks, Beaman and Haidle independently agree that cutting meat with stones is a substantially more complex and demanding task than fishing for termites. This indicates that early members of the genus *Homo* were engaging in technologically based tasks that were more demanding of working memory than those we see in chimpanzees today. So even if Wynn and Coolidge are correct that full-fledged, enhanced working memory arose only relatively recently, the process toward this may have started much longer ago.[49]

Modern cooking puts substantial demands on working memory. Consider cooking an apparently simple meal of buttermilk fried chicken, seasoned greens, and cornbread. Assume you start with all of the basic ingredients. Cooking the meal involves a large number of steps. For the chicken: cutting it up, placing it in buttermilk to marinate, removing it from the buttermilk after a certain length of time, seasoning flour, dredging the chicken in flour, heating oil for deep frying, and cooking the chicken piece by piece until done, making sure not to overcrowd the oil and lower its temperature too much. For cooking the greens: chopping them, heating up a large pot of water to boiling temperature, placing the greens, some salt pork, and seasonings in the pot, and simmering for an extended period of time until the greens reach a desired tenderness. For the cornbread: turning on an oven, mixing wet and dry ingredients in a bowl, pouring the mixture into a baking pan, placing it in the oven, and cooking until done. All three of these preparations should be conducted simultaneously.

To tally it up, this meal of fried chicken, greens, and cornbread requires the cook to make use of an impressive array of tools and techniques: two knives (one to cut up the chicken, another to chop the greens); three liquid media (buttermilk, water, and oil) used for substantially different purposes at different temperatures for different lengths of time; at least five cooking vessels (for marinating, boiling, deep frying, mixing, and baking) whose structural characteristics must match the varied tasks at hand; and a variety of other utensils that serve as extensions of the hand and arm to deal with the hot food or to serve it politely. In addition, the cook must assess when each item is finished cooking, which in each case requires monitoring a different kind of transformation of the original raw materials.

The working-memory capacity that allows a human to cook a meal of this complexity is truly beyond the capability of any other animal. But to give chimpanzees their due, fishing for termites is probably not the most working-memory-intensive food-related task they do. When chimpanzees (mostly males) hunt cooperatively, capture a prey item, share it among themselves and others, and in some cases trade meat for sexual access to females, this involves a complex series of actions that goes beyond termiting in terms of its demands on working memory. True, it does not involve tool use, but chimpanzee social behavior is more complex than their technological behavior.

It may be something of a moot point to identify when exactly humans or our ancestors achieved a "modern" level of working-memory capacity. The process whereby it happened should be more important than identifying arbitrary landmarks along that trajectory (although "thinking like a modern human" is a pretty important one, as far as arbitrary evolutionary landmarks go). Clearly, the concept of working memory helps us understand how we evolved and used the increased intelligence that gave us a cognitive advantage over the great apes and other, now extinct, hominins. If we want to explore the forces of natural selection that enhanced and shaped working memory, food preparation should be one of the most important to focus on. Increased complexity of food preparation makes for a more adaptable and flexible hominin. Given that food is embedded in a social network of sharing and reciprocity with friends and family, the working memory must be able to simultaneously manage the demands of technological, culinary, and social goals. When we look at the central place of food, including its preparation and distribution, in cultures throughout the world, we can be certain that achieving these goals was critical to evolutionary success at both biological and cultural levels.

Looking Forward to Remembering

We have all had the befuddling experience of striding purposefully into a room only to forget why we went there in the first place. But what was the nature of what we were forgetting? Did we have a memory to forget? Can something we never did actually be forgotten? Cognitive researchers call the kind of memory that involves remembering to do something in the future "prospective memory."[50] Often when people complain that they have a bad memory, they actually are referring to poor prospective memory. This kind of memory failure often provokes concern because it is so, well, memorable. What people are less aware of is the many times that their prospective memory does not fail them.

Prospective memory is really quite complex. It means keeping something in mind, yet not letting it interfere with all the things you need to do before you get to that task. When we consider the many things that we might plan to do in a day, we see that juggling these prospective memories is something we are all pretty good at. While many people make written to-do lists, most of us also have implicit mental to-do lists that help us keep prospective memories organized. For example, we have probably all had the experience of forgetting to do something after an event planned previous to it did not go as expected. The subsequent forgetting of the next activity suggests that a mental to-do list existed but was disrupted.

As you might expect, prospective memory calls into play several cortical regions of the brain.[51] Imagining a near future while at the same time putting it aside to do other things is not a trivial undertaking. People who have injuries (or lesions) to their frontal lobes can have all sorts of problems with planning and forward thinking—the executive functions. These can range from catastrophically poor decision making to absentmindedness. Absentmindedness is more or less equivalent to having a constitutionally poor prospective memory. Lesions near the very front of the frontal lobe (the frontal pole) are associated with deficits in prospective memory; this pattern is largely confirmed by functional imaging studies. Prospective memories do not seem to be stored in the frontal pole region, but it is a region critical for maintaining attention while doing another task. Other parts of the cortex are also activated when doing prospective memory tasks, creating a network that balances the cognitive demands of attention and intention.

Food acquisition and food preparation depend quite heavily on prospective memory. Cooking anything requires planning steps and anticipating the completion of the cooking process. If cooking with fire really is as ancient and as central to human evolution as Richard Wrangham hypothesizes that it is (i.e., older than 1 million years), it may have fostered natural selection of enhanced prospective memory.[52] There is a real potential risk in cooking food, in that if it is overcooked, its nutritional value can decrease or even be destroyed. This makes overcooked food a waste of all the effort that went into obtaining and cooking it. As anyone who grills knows, it is important to pay attention at certain times to what is going on with the fire. But as the saying goes, a watched pot never boils; furthermore, sitting around watching a pot boil, so to speak, may have been a luxury that our ancestors did not always have. There would always be an advantage in being able to attend to other things—children, wild animals, information being shared around the campfire—while cooking.

The ability to multitask, to attend to several things at once and maintain an internal to-do list, is theoretically one skill (among many others) that could have supported selection for increased brain size in human evolution, as it necessarily requires a network involving several parts of the cortex. The demands food and cooking make on prospective memory expand when an individual collects and prepares food not just for him- or herself but also for a partner, children, extended family, or a social group. The demands of several hungers must be anticipated and met. In other primates, planning about food acquisition is never more complex than a mother having to find enough food for herself and one dependent offspring. This kind of feeding depends on memory about food sources but really does not require prospective planning.

Looking at food and memory at a completely different level, cultural anthropologist David Sutton also uses the term *prospective memory*, but in a manner quite different from how it is used in cognitive psychology.[53] Sutton has spent several years studying cultural foodways on the Greek island of Kalymnos. Food is always embedded in culture, but Sutton has found that looking at food and culture from the perspective of memory is a way of linking the daily life of individuals with the historical traditions embodied in religion, ritual, and status. He defines prospective memory as "orienting people toward future memories that will be created in the consumption of food."[54] For example, when people in Kalymnos consider investing in a neotraditional outdoor oven to cook their Easter lamb, a strong incentive to build one comes from prospective memories of future Easter feasts and family celebrations.

Prospective memories associated with food are used on Kalymnos to mark the passing of the seasons, agricultural and fishing cycles, and dates on the religious calendar. Sutton emphasizes, however, that this does not mean that food is simply a mnemonic device that helps mark the passage of time. Kalymnian people have a real enthusiasm for foods as they come into season; they anticipate the coming abundance of one food or another and the change in diet this affords. Seasonal foods are also linked with fasting periods and feasts associated with the Orthodox Church. Prospective memories embody and express the enthusiasm and anticipation of these seasonal events. Sutton writes: "Religion and ritual are naturalized through the practices of everyday life, and vice versa. But equally important is the linking of past, present, and future in such practices."[55] These links are forged in part from prospective memories.

Cognitive psychologists Thomas Suddendorf and Michael Corballis argue that humans may be unique, and have a great adaptive advantage over other primates, in their ability to link past, present, and future.[56] We locate ourselves in a chronological narrative; Suddendorf and Corballis call the ability to move within this narrative "mental time traveling." Clearly, the human ability to imagine possible complex scenarios, tell stories, and plan for future actions has been part of what sets us apart cognitively from apes and some other hominins. Suddendorf suggests that this ability could be referred to as "episodic foresight," in contrast to episodic memory.[57] He argues that the two encompass distinct processes in the brain, although necessarily overlapping in some of their components.

The two kinds of prospective memory we have discussed here sandwich the concept of episodic foresight. The prospective memory of the mental to-do list is not necessarily episodic, but it relates to the ability to plan and imagine tasks and actions in a proximate and relatively immediate sense. In the cultural sense, prospective memory transcends episodic memory because it rests not in individual memories but in the collective memory of a culture. Yes, individuals carry and give voice to these prospective memories, but their shared content is wholly cultural. A prospective memory of traditional events can be formed without a basis in any individual personal experience.

Episodic foresight reflects part of the evolved biological capacity of the human brain. It is likely that it could have fully evolved only in a hominin species that already possessed some degree of human-like language and culture. Cultural prospective memories depend on episodic foresight, but in turn they demonstrate how the cultural can transcend the biological. I accept that the language I speak and think in is not my own, but I am biased to think that my memories are my own creation and belong just to me. The existence of cultural prospective memories supports the idea that there is a collective memory that shapes my individual memories. This is much more likely to happen with the things that concern us most. And on a day-to-day basis, from the moment we are born until the moment we die, there is nothing that concerns us more than food.

Food Feasts as Memory Feasts

In the previous chapter, we discussed the central role that feasts hold in human cultures and in human evolution. Another reason for the importance of feasts lies in the fact that they serve up not only an abundance of food but also an abundance of memories. Feasts are memorable events in the course of an individual's lifetime, but they also encompass the experiences of many others in previous generations who have shared in the ritual. Again, as with the cultural prospective memories, feasts are only partially mnemonic devices (although they do serve that purpose); they are themselves ritual from which history and memories are made.

In an overview of the anthropology of food and memory, Jon Holtzman identified several areas in which cultural memory is played out in relation to food.[58] Let me use his scheme to briefly consider that most memory-laden of traditional American feasts, the Thanksgiving dinner. Thanksgiving is a harvest feast that occurs in the fall (end of November) and features a large turkey as the centerpiece of a complex, multicourse meal. Although many cultures have feasts associated with harvest, the American Thanksgiving has become more than this: it is also a celebration of national history and identity.

Food and Sensuous Memory   An abundance of side dishes that vary in textures and seasonal flavors are served in support of the turkey. Beyond this, the Thanksgiving meal often is memorable for the spatial distribution of eaters, offering a visuospatial memory that reinforces other kinds of memories. The long table, immortalized by the likes of artist Norman Rockwell, hosts the largest dinner party of the year. The Thanksgiving meal is special not just because of the food eaten but also for the space it occupies in the typical household. If present, the children's table sets off a separate space and helps reinforce childhood memories of reduced status and the separation of the adult and child worlds.

Food and Ethnic Identity   Thanksgiving is all about being American, especially a certain kind of American. In a nation known for self-confidence and brashness, Thanksgiving is a time to show that Americans are not above giving thanks. At the same time, the displays of abundance remind everyone that success is not something to be ashamed of, especially if it is shared with others. It also reinforces the notion of America as a melting pot—preparing a Thanksgiving meal and eating it in one's own home marks the Americanization of the immigrant as much as receiving a green card or citizenship does. Many immigrant families remember their first Thanksgiving as a significant milestone in the adoption of their new ethnicity.

The Gastronomic Memory of Diaspora   The Thanksgiving meal is a celebration of internal migration in the context of immigration. The "homeland" in this case is located in space and time in seventeenth-century New England. All Americans become, in a sense, New England settlers through this meal. At the same time—and this is part of the universal appeal and success of Thanksgiving among America's diverse immigrant ethnicities—it effectively highlights the universal experience of immigration. A reminder of one migration is clearly an effective mnemonic for remembering others.

Gustatory Nostalgia, Experienced and Invented: Food, Nationalism, and Invented Traditions   If Thanksgiving did not already exist, someone would have to invent it. Janet Siskind's analysis of the invention of Thanksgiving, a long and complex process involving the blending of several traditions, makes clear that there was a need in the Unites States for such a unifying story of national identity. As she writes: "Thanksgiving powerfully shapes a sense of nationality to the emotions of homecoming. The joys and tensions, pleasures and pains of family life are activated in the preparations and joined participation of the feast."[59] Nostalgia and nationalism are linked by a simple story of national founding and unity. The memory of this nationalistic "event" exploits the memory-heightening power of the emotion of personal relationships and the procedural memories of the preparation of the feast.

Food, Gender, and the Agents of Memory The traditional Thanksgiving meal and its preparation reinforce stereotypical gender roles, with women in charge of the preparation of the meal while the men lead the ritual of its serving, carving the turkey at the head of the table. Even with some breakdown of traditional gender roles concerning the preparation and serving of the Thanksgiving meal, it still provides opportunities to display aspects of gender. For example, the competitive eating aspect of the Thanksgiving meal is probably still largely seen as a more masculine rather than feminine reflection of American society. Overconsumption is common regardless of gender, but the display of it is more male, especially for food consumption. The image of males sitting around the TV after the meal, bellies distended and belts loosened, is a memorable one.

Food as the Marker of Epochal Transformations Thanksgiving marks many transformations: from the Old World to the New World, from failure to success, from childhood to adulthood. Birthdays reflect what may be the most profound transformation of all, from non-life to life. Interestingly, Thanksgiving is not a birthday celebration; the United States has the Fourth of July or perhaps Columbus Day for that. Rather, Thanksgiving marks the passage out of infancy, a dangerous and vulnerable period that must be survived if adulthood is to be reached. Of course, individuals have no memories of their infancy; the early formative years go by in a blur. Thanksgiving provides a nice memory of a secure infancy, an almost cute representation of a nation, not unlike a snapshot of a baby in a warm family setting.

Rituals of Remembering and Forgetting through Food   Almost all feasts are memory feasts, but the selection of what to honor in a feast also involves decisions about what not to honor, about what is better left forgotten. The traditional first Thanksgiving is presented as an example of cooperation between Native Americans and Pilgrim settlers. As such, it represents a portrayal of relations between Native Americans and European immigrants that is diametrically opposed to the real character of that relationship as reflected over centuries of conflict and ultimate dominance by the colonizing invaders. Collective amnesia may be more dangerous and powerful than collective memory.

Maurice Halbwachs was one of the pioneers in thinking about and defining the notion of collective memory. I am quite taken by something he wrote: "Often we deem ourselves originators of thoughts and ideas, feelings and passions, actually inspired by some group. Our agreement with those about us is so complete that we vibrate in unison, ignorant of the real source of those vibrations."[60] Both eating and remembering are individual activities. I eat, and I remember; you eat, and you remember. But somewhere between and around us are various "vibrational" forces that make my memories about food converge with yours.

Our shared physiology and biological history shape the various kinds of memories that are relevant to food and eating. There is reason to believe that food is a privileged target of memory, given that the hippocampus is sensitive to an array of circulating hormones. But memories surrounding food go beyond the physiological. A common culture infuses food with shared meaning and emotional resonance, and our minds construct food memories while being influenced by these cultural factors. Culture in turn changes the selection environment for various cognitive abilities, leading to the evolution of enhanced working memory and episodic foresight. Food preparation and acquisition techniques both drive and benefit from these evolved enhancements.

Nothing defines the present like remembering the past. It seems to me that one definition of a happy person would be someone for whom the good memories outweigh the bad. For those fortunate enough not to be living under conditions of food scarcity, food, sometimes in its simplest form, provides a potentially ready source of good memories. By keeping in mind the various forces, individual and collective, that shape our food memories, perhaps we can do a better job of creating memories about good food and good times and putting aside the bad.

6

CATEGORIES: GOOD FOOD, BAD FOOD, YES FOOD, NO FOOD

Soft-boiled eggs

* * *

Beef liver aux fines herbes

* * *

Cold meat

* * *

Swiss cheese

—AUGUSTE ESCOFFIER's menu for September 1, 1870,
   the day of the Battle of Sevigny

IN THE LATE NINETEENTH CENTURY, Auguste Escoffier become
the embodiment of the complexity and sophistication associated
with classical French cuisine. He promoted and popularized this
vision of French food primarily through the kitchens and dining rooms of the Ritz hotel chain. Many years before, however,
the young Auguste Escoffier served as an army cook during the
Franco-Prussian War (1870–71). In his memoir, Escoffier describes in great detail the lengths he went to in order to make
sure that his men, and especially his officers, were able to enjoy as
high a standard of cuisine as possible under the trying circum-

stances. In addition to the basic military stores of tinned meat and fish, Escoffier raided the countryside and village markets for sources of fresh meat, poultry, eggs, vegetables, and herbs. Suckling pigs were obtained and quickly converted into pâtés, which he called *les pâtés du siège de Metz* after the pivotal battle of the war.[1]

Cooking while having to cope with the privations of war could not have been easy for Escoffier. Yet it is clear that he managed to maintain his sense of what a meal should be, despite the conditions. The menu above, one of several battle meal menus he provided in his memoir, is a manifestation of the mental template for "meal" that Escoffier carried in his mind. The reality was a meal composed of eggs obtained locally, leftover beef and liver from the previous night's dinner, herbs collected on the go, and some cheese. More than likely it was served not elegantly, in courses, but on a single tin plate. Yet it is clear that for Escoffier, the four separate components of the meal could be categorized into courses.

For Escoffier, this exercise in categorization was part of what made this collection of edibles a meal. In recording this and other wartime meals in his memoir in formal menu form, Escoffier raised them to a level higher than simple military grub. He put these humble repasts on equal footing with the extraordinarily complicated meals he later prepared for royalty and other celebrities. Escoffier retrospectively fashioned these simple menus to show that he cared about and put great thought and effort into his wartime cooking. Although he was limited by raw materials and primitive conditions, he did not abandon his principles or his training. He did not stray from his mental template of a meal.

One of the basic distinctions that all humans employ is that between food and non-food. No one eats everything in the environment that can be consumed and digested by a human. It seems to me that Escoffier took this distinction one level higher: he made it clear that he provided *food* for his men, not simply *fuel*. Escoffier's battlefield cooking endeavored to go beyond mere sustenance, despite the wartime conditions. Escoffier worked to turn the results of his foraging into a cultural product, a creative expression of his mind, rather than leave them as a collection of nutritive substances. Central to this act of reclassification was classification itself, placing the different foods into their ordered spots in the menu.

Food is certainly one of the most important things in the environment that needs to be classified, and food versus non-food is perhaps the most basic categorization of all. We are all aware of cultural dietary prohibitions, such as the Jewish or Islamic ban on eating pork. The origins of these sorts of food prohibitions have been much discussed in anthropological circles, and what is or is not eaten—what is or is not food—contributes mightily to cultural identity.[2] We can also develop our own idiosyncratic food prohibitions—for example, if we have an aversive reaction to eating something and refuse to eat it again. That sort of aversion goes beyond simply not liking something—it is no longer food but a toxin. But one culture's toxin may be another's food. Substances commonly labeled as "drugs" in one culture may be considered foods in another.[3]

Cultural prohibitions against eating certain foods are, of course, vitally important in some contexts. But in the sense that they reflect cultural indoctrination and learning as much as food habits, they do not give us too many insights into the cognitive nature of human food categories. On the opposite end of the spectrum, the kind of aversion that develops upon eating a food that induces vomiting is also not really an interesting example of how the mind makes and uses categories: such an item more or less gets classified as non-food without too much thought being involved in the process. Naturally, to a Jew who keeps kosher, prohibited foods would be in a different category from foods that are avoided due to an aversion reaction.

In this chapter, I am more interested in exploring how we classify foods rather than non-foods. Humans like to classify things in the natural world. More than this, how people see and interact with objects in their environment depends on classification and categories. What those categories are is less important that the fact that categories are made and adhered to. An omnivorous mind puts a premium on the brain's ability to order the objects of an arbitrary world (arbitrary, that is, from the perspective of any particular actor). Food is not necessarily a privileged item in our brain in terms of classification, but we cannot understand how the omnivorous mind thinks food unless we understand how the brain classifies anything and everything.

Turkeys versus Cassowaries: Generic and Other Kinds of Species

Biologists have their own, formal way of naming all of the plants and animals in the natural world, living and extinct. This system (biological taxonomy) has agreed-upon rules and procedures, which are recognized internationally and enforced consensually (there are no taxonomic police to hunt down rogue classifiers). The taxonomic system biologists now use dates back to a scheme introduced by the Swedish biologist Carl Linnaeus in the eighteenth century. The scheme is hierarchical, with smaller or more limited categories nested within larger and more inclusive ones. For example, the order Primates is in the class Mammalia, which in turn is nested within the subphylum Vertebrata. Working down from primates, we move through various taxonomic levels until we reach the species level. Following Linnaeus, species names are usually expressed using two terms, the first indicating the genus and the second the actual species designation. Thus, for example, humans are *Homo sapiens*, chimpanzees are *Pan troglodytes*, and gorillas are *Gorilla gorilla*.

The biological system of classification has been around for a long time, and most biologists are comfortable with it. For a growing number of scientists, however, that is not a good enough reason to keep using it. Linnean taxonomy reflects an eighteenth-century endeavor to classify all of God's creation, which at the time was thought to involve at most tens of thousands of species; in the twenty-first century, however, we know that there are millions of species, whose classification, most scientists believe, should reflect evolutionary relationships as accurately as possible.[4] The old system of biological classification does not systematically make use of genetic and evolutionary information derived from new molecular technologies. Replacements for Linnean taxonomy have been proposed, but people become very attached to the names and categories they are used to. Since for most biological research questions classification is a side issue, not something that's directly relevant, many biologists can work around or do not have to deal with the limits of this old system.

Biologists generally agree that the species level of classification is critically important to understanding life on this planet. Volumes have been written on how species should be defined and how the species concept should be applied. The importance of species in scientific circles reflects to some extent the importance of species-like categories seen in all human cultures. Ethnobiology, or folk biology, is the study of how different cultures organize the natural world.[5] There is much variation in how cultures do this, as we'll see a little later in this chapter, but all classification schemes depend on the identification of "generic species." Organisms placed into a single generic species (and labeled with a common term) are assumed to share an essence of some kind. People in all cultures expect to be able to place all organisms in one or another generic species.[6] In other words, they expect to be able to classify anything, although they may not bother to classify everything in their environment.

Biological organisms are just one of the things that humans classify in the natural world. It is reasonable to assume that folk biological classification probably shares cognitive processes with the classification of other objects or even concepts. People create generic categories, defined by certain essential properties, and try to slot things that are important to them into these categories. Food—how it is obtained, prepared, and eaten—is one of these important things.

In the 1960s, anthropologist Ralph Bulmer undertook a long-term multidisciplinary project to understand how the Karam (or Kalam) people of the highlands of New Guinea organize the natural world around them.[7] The Karam are horticulturalists living between 5,000 and 8,600 feet above sea level on land that encompasses both open grassland and forest. They raise pigs and several crops, supplementing their diet by hunting and gathering. Bulmer's classic and pioneering study of Karam folk biology serves to illustrate how people classify things based on essential features (either implicit or explicit) and how these essential features may vary widely depending on wider cultural concerns.

In studying the zoological taxonomy of the Karam, Bulmer asked himself why the cassowary is not a bird, at least according to the Karam.[8] Now, in scientific taxonomy, there is no doubt that the cassowary is a bird. It is a member of a somewhat curious category of flightless birds, some very large, known as the ratites. These "feathered dinosaurs" (a term that reflects another classificatory ambiguity) include birds such the ostrich, rhea, and kiwi, and the extinct moas and elephant birds, all found in the Southern Hemisphere on lands that once made up the supercontinent Gondwana. The family Casuariidae includes three species of cassowary and the one surviving emu species, distributed in Australia, New Guinea, and some of the other islands of Melanesia. All of the cassowaries are distinguished by a high bony crown on the top of their skulls. They live in forests and are generally reclusive; they can be aggressive when confronted, and their large, clawed feet can be dangerous to attackers. They have a reputation for being a man-killer, although this seems to be greatly exaggerated.

Unlike scientific ornithologists and neighboring ethnic groups in the New Guinea highlands, the Karam do not think that the cassowary is a bird. Bulmer noted several pretty self-evident reasons why this might be the case: the cassowary does not fly; it has only rudimentary wings, which contain no feathers but just sharp quills; it is much bigger than any other bird the Karam are familiar with; and its bones are very strong and robust, unlike the bones of other birds. When he naively asked Karam people why cassowaries were not birds, they came up with two more features: cassowaries have no feathers but "hair," and they have very tiny brains compared to the size of their skulls. Bulmer agreed that the feather form of the cassowary is hair-like, and therefore a reasonable anatomical feature to use to distinguish them from other birds. But he was surprised by the brain size comment, in that he had no other evidence that cranial structure and bone-to-brain ratios were relevant in Karam taxonomy. Upon reflection, he realized that it is indeed true that, given the large bony crowns on the tops of their heads, cassowary brains are indeed very small compared to the size of their skulls.

Bulmer argued that while all of these anatomical features could have led the Karam to not classify the cassowary among the birds, anatomical features alone did not really make for a sufficient explanation of this classification. Among the Karam, hunting of the cassowary is a culturally distinct and highly regulated activity in comparison to hunting of other animals. Men preparing to hunt cassowaries have to practice avoidance, meaning that they cannot use everyday vocabulary to refer to various objects and activities. In dispatching a cassowary, no blood is to be shed; they have to be caught in a snare or killed with a blunt instrument. The hunter who kills the cassowary has to eat its heart. People who kill or eat a cassowary become ritually dangerous and are not allowed to plant or be near growing taro crops for one month. With only one ceremonial exception, all cooking and eating of cassowary should occur in the forest or at the forest edge, and nothing used in their preparation should be brought into settlements near taro gardens. Finally, no live cassowaries, whether adults or chicks, can be brought into the home or garden. This prohibition stands in stark contrast to the practices of other ethnic groups who routinely raise cassowary chicks in their settlements for trade or food. The practical upshot of all these rules and regulations is that cassowary hunting among the Karam winds up involving only one or two male hunters at a time, who do not so much hunt the cassowary as engage in a duel with it.

In exploring the place of the cassowary in Karam culture, Bulmer discovered that the Karam put a strong emphasis on the physical and ritual distinction between the cultivated world and the forest world. This distinction is reinforced in myth and a variety of cultural practices. Based on the ritual structure of the

Karam landscape and numerous comments and observations from his informants, Bulmer came to the conclusion that cassowary is not a bird but a "quasi-human" of the forest, "metaphorical cross-cousins" to the Karam:

> To slay these creatures is in some sense equated with killing a man. To kill a man makes one *asŋ*, ritually dangerous, in just the same way as to kill a cassowary or a dog. Killing a cassowary is like committing homicide in yet another respect. If one kills a cassowary, one must eat its heart *(md-magl)*. If one kills a human being one doesn't eat one's victim's actual heart, but as soon as possible one kills and cooks a pig and eats its heart instead. . . . When one eats the heart of a cassowary, people say, one ensures that its spirit goes back to the forest, and will not prevent one from killing more cassowaries in the future.[9]

As anyone familiar with the rudimentary cosmology of the American Thanksgiving turkey can probably understand, sometimes a bird hunted and eaten is not simply dinner, but something altogether more meaningful (though it can still satisfy an appetite). The human relationship with the natural world—and for most of human history food was natural, as opposed to the highly processed commercial products common in developed societies today—is mediated by all sorts of factors or "biophilia values," as Stephen Kellert describes them.[10] Building on Edward O. Wilson's biophilia hypothesis, which holds that humans have an evolved bond with nature, Kellert identifies ten values that influence the human relationship with the natural environment. Included among these are utilitarian, ecologistic-scientific, aesthetic, symbolic, and moralistic concerns. Cultures vary in the weight they give to these various concerns, and what a culture does or does not emphasize helps to determine that culture's relationship with the objects and beings in the surrounding world.

A culture's relationship with nature colors all aspects of how its members categorize natural objects. When Ralph Bulmer went to highland New Guinea and saw a cassowary, he categorized it as a bird based on Western scientific and utilitarian principles; in addition, there was probably an aesthetic component in Bulmer's own classification of the cassowary, as he was very much interested in birds. The Karam, too, classified the cassowary with both utilitarian and aesthetic values in mind (it is eaten as food, its bones and claws can be used as tools, and its feathers can be used for headdresses), but it is clear that there was also a strong symbolic and moralistic component in how they viewed the cassowary. Similar to the way that a turkey is not just a bird to Americans, a cassowary cannot be a bird for the Karam.

For both Americans and the Karam, underlying the folk biological classification of turkeys and cassowaries is an implicit concept of a generic species. The essential features of these generic species may not typically be identified, discussed, or acknowledged, but they nonetheless contribute to cultural classification. Shared knowledge can be explicit or implicit. One way cultures organize and share implicit knowledge is by using "natural" classification schemes, or at least schemes that seem natural to them. So when I made a classificatory judgment about Escoffier's cooking for his soldiers, differentiating between food and fuel, I tapped into a cultural classificatory scheme that already existed. The "rules" for this classification are obviously complex but implicit, at least to me, until I thought about why I made such a distinction.

Is it Pizza or Not? Categories and Classifications in the Brain

As we go through the cultural classification of the cassowary and the turkey, it is apparent that we are dealing with different levels of cognitive processing. On one hand, there are numerous cultural factors, collective and individual memories, specific learning experiences, and so on that all might shape how an individual or culture classifies these animals. On the other hand, if I show a picture of a cassowary to a Karam person and to a birdwatcher familiar with scientific taxonomy, and I ask, "Is this a bird?" the Karam person might quickly answer no and the birdwatcher yes. The object "cassowary" resides somehow (not necessarily somewhere) in the brain, where it can be quickly accessed in order to answer a question concerning its place in the universe. The quick recall of an object slotted into an existing classification is a skill needed numerous times over the course of a normal day, which suggests that applying categories is a streamlined cognitive process even if defining them is not.

When the anthropologist asks the "why" question, things get complicated as we sift through the many possible cultural influences on classification. Why the cassowary is a bird would take just as long to explain as why it is not. Psychologists and other cognitive scientists often ask the "how" question—how do people make, learn, and use categories? In order to answer "how" questions, researchers must develop experiments and tasks that simplify categorization and classification so that these phenomena can be studied in the laboratory and the relevant variables controlled.

In the field of psychology, there has been a vast amount of research on the cognitive basis of categorization.[11] This is understandable given that categorization is one of the human mind's

most important tools. Several tests have been designed to tease out
the various psychological components of human category learning. Although these can seem to be somewhat artificial in the sense
that they employ playing cards with random or nonrandom dots
and other such stimuli, it is important to keep in mind that in
some sense every human classification system is artificial—order
imposed by our minds on a fundamentally disordered world.

Cognitive psychologists have identified at least four major strategies by which people learn perceptual categories—"collection[s]
of objects belonging to the same group."[12] One fascinating and
informative finding is that patients with brain diseases, such as
Parkinson's disease, can sometimes do well with one category-
learning strategy but not so well in others. This suggests that
categories can be learned via different and distinct pathways in the
brain, and what we perceive to be a single category or kind of
category can be learned or applied via different cognitive routes.

Rule-based learning is perhaps the most intuitive form of category learning. A rule is developed based on an explicit reasoning
process and then is applied to different objects to see if they belong to the category based on that rule. The rules are explicit in
the sense that the categorizer could identify them and state them
to another observer. For example, the category "hamburger" could
be defined by the rule "a sandwich made from ground beef on a
cut roll." You can put cheese on it and call it a cheeseburger, or
add any number of other toppings, but it would still be a hamburger. In New Zealand, where I lived for many years, a traditional hamburger has a fried egg and sliced beets on it. This was
confusing to my American hamburger norms, but there was no
doubt in my mind that it was still a hamburger, albeit a somewhat
odd one. The basic rule generally takes precedence over the variation introduced by toppings. On the other hand, if the ground

beef is replaced with another form of ground meat (e.g., lamb, pork, or turkey), I think that many people would not accept this as a hamburger, although one could argue that it is categorically related to the "true" hamburger. For various reasons, some historical and some cultural, the meat in the hamburger is critical to defining that category, while the highly variable toppings are not. We can have multiple rules to define a category, although there are limits to how many the human mind can handle at a time. For a formal category system, such as scientific taxonomy, the rules can be more numerous, of course.

Another set of category learning tests is based on prototype and/or exemplar learning. There is actually some debate in the field over how big a difference there is between prototype and exemplar learning.[13] A category can be defined by a prototype, against which all other objects are tested. If an object does not deviate too much from the prototype, then it can be classified in the category defined by the prototype. In contrast, a category can be defined or learned by examining exemplars of that category; the members of the category all share attributes that warrant their grouping, but no single example is held to be the prototypical example. Based on knowledge of the category via exemplars, it is possible to decide if a novel item should be included in it or not.

Although I think the hamburger category is best understood in terms of a primary rule, I can see that one could also easily create it from its exemplars. I am not sure that a general hamburger prototype exists. But whatever the psychological basis for one type of categorization versus another, individuals differ in the dominant mode they use for defining a particular category. For example, imagine that a person's first exposure to pizza is Chicago deep-dish pizza. Over time, other pizzas are experienced, and gradually a category "pizza" is developed via these exemplars. It is a fairly inclusive category, including thin- and thick-crust, deep-dish, Hawaiian-style, and even abominations such as whole-wheat (I spent many of my formative years in Berkeley, and I can't tell you how many times I attended a function only to be disappointed by the serving of pizza with a whole-wheat crust). In contrast, imagine those eaters who, for whatever reason, are obsessed with "authenticity." For them, "real" pizza is the kind that developed around Naples, Italy, evolving from Roman flatbreads and reaching its ultimate form no later than the middle of the nineteenth century.[14] It has a thin crust, is cooked in a very hot wood-fired oven, and is not overloaded with toppings. All other pizzas are recognized as such only as distortions of this Neapolitan prototype.

The category "pizza" will look pretty much the same whether it is filled by someone who is more exemplar-oriented or someone who is more prototype-oriented. But I suggest that enthusiastic foodies can be divided into these two basic types, exemplar eaters and prototype eaters (most people, of course, wind up somewhere in the middle, sometimes one and sometimes the other depending on the food and circumstances). The exemplar eater enjoys variations on a theme, new versions of old dishes, and takes a more relativistic approach in assessing the quality of food. The prototype eater not only looks to the original version of a dish as setting a standard but also seeks new prototypes in the form of the "best" version, in effect to provide the category with a reference exemplar. Of course, this does not preclude the prototype eater from enjoying different versions of a dish; indeed, part of the fun in finding the best version is sampling all the near-best versions along the way. But eaters who need to find the best barbecued pulled pork, French baguette, sushi, or burrito in their communities are prototype seekers. All eaters can benefit from their efforts, especially if they share their results via Internet reviews.

For both rule-based and prototype/exemplar categorization, it should be possible to communicate or declare the features, rules, or strategies upon which the categorizations are based. Psychologist Gregory Ashby and his colleagues also emphasize that in the same way that we have explicit and implicit memory formation, there are probably also categories that are learned or used in an implicit manner, such that it would be difficult for the categorizer to state clearly all of the factors that go into placing items into one or another category.[15] One such implicit form is called information-integration category formation. In this context, categorization occurs only after there is integration of at least two (and sometimes many more) different components of whatever is being categorized.

An example of information-integration categorization occurs when an experienced chef cooks a piece of meat to a desired level of doneness. The conventional levels of doneness form a pretty tight categorical scale: bleu (surface cooked but just warm inside), rare, medium-rare, medium, medium-well, and well done. It's relatively easy to place meat into the right level of doneness category if one can cut a bite, examine its color, and taste it. Doneness can also be measured using a thermometer, although that can be difficult with thin cuts of meat; it is much easier with a thick roast. But we expect a good chef to be able to cook to these levels of doneness without resorting to cutting into and sampling the meat, or even using a thermometer. In order to produce the desired level of doneness, the chef must take into account the thickness of the cut, the kind of meat, its fat content, the heat transference properties of the cooking surface, and the actual amount of heat being produced. The chef relies on vision, touch, smell, and even an intuitive sense of time to make judgments about which category the meat is in at any given time during the cooking process. Large commercial kitchens have various ways to standardize this process (uniform cuts of meat, consistent grill temperatures), but a real chef would have a hard time conveying exactly how he or she knows when the meat is done. As the food scientist Harold McGee says, "The best instruments for monitoring the doneness of meat remain the cook's eye and finger."[16] What this really means is that experience counts, and that assessing doneness is best accomplished via an implicit process (in the chef's brain) rather than an explicit one.

One final category learning strategy is another more or less implicit one called the "weather prediction" strategy, after the name of the psychological test used to investigate this sort of categorization.[17] Basically, there are some category schemes that we use that are inherently probabilistic: objects can be classified into categories but there cannot be certainty that the categorization is correct. This is not to say that the classification is random; rather, each decision is made only with a probabilistic estimate of being correct. This sounds quite odd, but consider this hypothetical situation. Imagine that you are confronted with twenty foods that you have never eaten. Your task is to sort them into four categories: foods that you are very likely to enjoy, foods that you are moderately likely to enjoy, foods that you are moderately likely to dislike, and foods that you are very likely to dislike. Based on your own, implicit knowledge of your food preferences and dislikes, this should not be a very hard task. You will be able to sort them definitively into these categories; however, since the sorting can only be based on a probabilistic estimate of the likelihood of enjoying the foods, there is no way of knowing if any specific food is correctly categorized if you haven't tasted it. This is very much like the real-world experience of going to an unfamiliar restaurant, especially a high-end one with a changing and very creative menu ("Hmm, I like poached quail eggs and I like compressed watermelon, but would I like them together in a modern take on the traditional New Mexican sopapilla?").

The human brain imposes categories and classifications on a vast array of real-world and even imaginary phenomena by using a range of diverse strategies and a variety of sensory inputs. It should come as no surprise, then, that research on the neuroanatomical basis of categorization indicates that multiple brain regions and pathways are involved. As Carol Seger and Earl Miller write, "The brain does not have one single 'categorization area.' Categories are represented in a distributed fashion across the brain, and multiple neural systems are involved. . . . Categorization tasks are not process-pure: Multiple systems may be recruited to solve any categorization problem."[18] Regions of the brain that are implicated in categorization include the higher-level visual association cortex of the inferior temporal lobe, which is especially important for shape-based classification of things such as faces; similar association areas for other sensory modalities (smell, hearing, touch) are probably involved in category processing as well. The prefrontal cortex is active in abstract, rule-based distinctions. The parietal cortex links category discrimination to visuospatial processing. The premotor and motor cortices of the frontal lobe come into play when the outcome of categorization activities requires action. The learning and memory functions of the medial temporal lobe and hippocampus are critical for category learning and for cross-referencing among different categories. Forms of implicit categorization, such as those in which the category can only be determined via trial and error, depend on networks linking the cortex and subcortical structures, such as the basal ganglia.

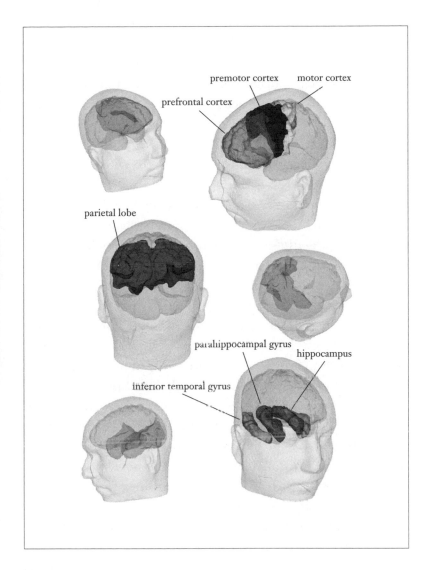

All of these diverse functional regions of the brain can come into play during categorization tasks, depending on the nature of the task and what is being categorized.

The ability to categorize and to base decisions and actions on categories has been strongly selected for in human evolution and in the evolution of other animals. Categorization abilities have been shaped by natural selection in the context of the overall behavior and ecology of species. For example, all social primates need to be able to classify facial expressions for their emotional content in order to function within the social group; functional imaging demonstrates that facial processing in humans draws on a network involving numerous brain regions distributed throughout the cerebral cortex, limbic system, and basal ganglia.[19] Dogs can be trained to classify odors based on any number of criteria that humans find useful; this ability probably builds on implicit categories that dogs use when they navigate through their own olfactory universe. In humans, language moves categorization from the implicit to the explicit and declarative level, as it does with memory.

Animals have to be able to separate (based on implicit classifications) food items from non-food items, not only for the sake of safety but also to avoid wasting energy obtaining non-nutritive substances. One likely reason that food categorization is so important to people is that no matter how complex, removed from the "natural" world, or vested in some higher ideology it may be, it is ultimately driven by the very fundamental and ultimate need to distinguish food from non-food. The emotional drives underlying food-seeking behavior and the cognitive reward systems that come into play when appropriate foods are obtained imbue decisions and ideas about food categories with an urgency that may be lacking for other aspects of the environment. Thus, for some people anyway, whether or not Chicago-style deep-dish is "real" pizza matters—a lot.

Why Do Diets Need Names or Shapes?

When people change their diets to lose weight, they often like to adopt one that has a name. The Atkins diet, the Zone, the South Beach diet, the Paleolithic Prescription—these are just a few of the literally hundreds of diet names that have been used over the years. A diet with a name may suggest many things—that it has been tested, that someone or some organization is willing to stand by it, that it reflects the temporary status of a weight-loss phase in contrast to a more normal, everyday diet. The named diet also indicates the change in status—the change in category—of the person embarking on the diet: once just an eater, now a dieter. Advertising this change in status, both to oneself and to others, undoubtedly helps psychologically reinforce the difficult task of losing weight.

People change diets and adopt new ones for reasons other than losing weight, of course. Many of these diets are what I would call "diets of enlightenment." For example, vegetarianism and veganism could be considered diets of enlightenment in cultures where meat-eating is common. Locavorism would be another diet of the enlightenment. There are many paths to the various kinds of enlightenment, and dietary enlightenment is no different. One can have an aha moment, or as Jonathan Safran Foer, in his extended argument for vegetarianism, *Eating Animals*, describes it after learning as a nine-year-old that the chicken he was eating was once actually a chicken, one of those *"how-in-the-world-could-I-have-never-thought-of-that-before-and-why-on-earth-didn't-someone-tell-me? moments."*[20] In contrast, enlightenment can come more slowly, after a period of education and reflection— say, someone becoming a vegetarian after reading Foer's book and considering its argument. Religious diets are also diets of enlightenment, although the nature of the food in the diet is typically a secondary concern.

Abandoning the metaphorical darkness and seeing the light can be a very powerful inducement to change. Part of the power in the dietary context stems from the fact that normative diets are not often named or labeled. One might say, "the food I grew up eating," "the way my parents fed me," or "how I used to eat." Things that are unnamed or unlabeled exist in a kind of cognitive darkness in the mind. Implicit knowledge is real and useful, but it exists below the radar of consciousness unless some effort is devoted to bringing it out. Tagging an object or concept with a label helps our minds retrieve it. Similarly, named weight loss diets are diets of enlightenment, in the sense that they help bring implicit eating habits into the cognitive light of day.

Although we may make use of implicit categories, explicit or declarative classification is perhaps what really separates us humans from other animals. Language is central to this process of classification, since explicit categorization requires labeling of objects and ideas. Humans have an impressive capacity for words. The average English-speaking high school graduate knows about 40,000 words; if we include place names, personal names, and various other idiomatic expressions, the count would be considerably higher.[21] The *Oxford English Dictionary* lists 171,476 words in current use and 47,156 obsolete words.[22] More than half of all English words are nouns, a quarter are adjectives, a seventh are verbs, and the rest various other parts of speech, such as prepositions. Thus three-fourths of all words in English are used to label or modify an "entity, quality, state, action, or concept" (from the definition for *noun* from Merriam-Webster's online dictionary). Other languages are also likely to have a disproportionate share of their lexicon devoted to nouns and adjectives.

The tens of thousands of words that humans can easily master far outstrips the ability of the great apes to learn symbols.[23] Even the most proficient of the apes who have learned sign language of some kind can only use in the neighborhood of hundreds of signs (which is still reasonably impressive). Some anthropologists and linguists argue that given our huge lexicon, humans have invented ways, essentially grammatical rules, to associate objects and actions.[24] The beginnings of grammar may be traceable to the fact that in order for a spoken string of words to be of any use, conventional rules signifying parts of speech *must* be developed or else there is no way for one speaker to understand another. In this view, grammar constitutes the overt rules people use in order to make relationships among symbolic representations of objects and ideas (i.e., words) understandable to others. There must also be covert rules for relating words that exist within the mind of each individual. When psychologists study category learning, they bring some of these covert rules to light. These grammatical rules are representations about the ways that people conceptualize the world.

The notion that grammar emerges from a generally large brain with an increased capacity to store and process words is more or less counter to the idea (derived from Noam Chomsky's writings in the 1950s and 1960s) that we have an innate or deep grammar that develops when we are children. I say "more or less" because even though researchers using computer modeling, information about other forms of animal communication (including that of sign-language-using apes), functional neuroimaging, and the variation in grammatical structure of the world's known languages are casting doubt on the deep-grammar view, we still cannot reject the idea that language has a privileged place in the functional anatomy of the brain and in human evolution.[25] Whatever this cognitive core of language might be, it could still be defined as a deep structure if not a deep grammar.

This digression into the evolution of language serves to make the point that words and labels are important and reflect how people relate to their environments. But not everything important is given a word label. All people use grammatical rules and the parts of speech, but actually labeling these things is an academic exercise. I think a similar situation exists for what we might call the normative diet. It is so assimilated, so thoroughly a part of a person's being, that only when it is being challenged by or contrasted to another way of eating does it need to be explicitly considered as a whole entity. In other words, we have a deep cognitive component to our mental view of food and diet. In keeping with the emergent-grammar view of language, I look at this cognitive aspect as something that necessarily develops as an efficient way to deal with the (food-related) complexity of the world around us. An important part of this view of food is an implicit understanding of food categories and classification.

The deep, normative view of food may be the reason that governments have such a hard time improving the diets of their constituents via official recommendations. One of the most common ways they try to do this is by developing an "official" graphic representation of the ideal or recommended diet. This recommended diet tends to reflect the international nutritional science community's current consensus on what people should be eating to stay healthy. In their survey of food guides from a dozen countries, James Painter and his colleagues note that "the core recommendation for individuals to consume large amounts of grains, vegetables, and fruits with moderate intake of meat, milk, and dairy products was consistent in all the international food guide illustrations."[26] In terms of their content and categories, the food

guides are surprisingly devoid of indigenous or traditional content, strongly and somewhat blandly reflecting nutritional science's view of things. It is also important to note that these food
guides are often very much the product of committees, reflecting
the interests of the medical and public health communities as
well as the financial concerns of the food industry.[27]

The shapes of the guides vary more than the content. Americans are familiar with the food pyramid, a format also adopted
by nutrition groups in the Philippines and Puerto Rico.[28] A five-
stepped pagoda is used in China and Korea. Various circles are
used to represent portions in Australia, Germany, Portugal,
and Sweden; the circles are rendered as plates in Great Britain and
Mexico (in the summer of 2011, the U.S. government announced
it was abandoning the pyramid in favor of the plate). The pyra-
mids and pagodas are designed so that the base of the structure
represents the things we should consume the most of in a good
diet, while the things we should eat less of are represented by
the reducing size of the pyramid or pagoda as it goes upward. The
dietary circles are basically elaborated pie graphs (although pie
of any kind, sweet or savory, is, alas, not typically included in
the recommended diet).

The most deviant illustration may be the one from Canada,
which is shaped like half a rainbow. Foods that are to be eaten the
most are represented in the outer rings, while the smaller inner
rings contain the foods that should be eaten more sparingly. But
Canadians are supposed to eat all of the colors of the rainbow.
The official food illustration for Japan is also an outlier in that it
takes the form of a spinning top, a more dynamic and inverted
version of the food pyramid.[29] The top is a traditional Japanese toy
and thus evokes Japanese culture. The spinning of the top represents physical activity, and the fact that it is balancing reflects a

balanced diet. Even the stem of the top represents something: water and tea. Otherwise, the categories in the Japanese spinning top are similar to those used in other countries.

The main titles for these food illustrations are all rather determinedly low-key and nonthreatening. In Japan, the heading asks, "Do you have a well-balanced diet?" The Australian circle exhorts, "Enjoy a variety of foods every day." The Swedish circle is helpfully titled "The Food Circle," and below that it says, "Something from each group everyday—Choose fiber-rich and low-fat products." No "Secret diet of the Hollywood stars!" or "Eat this way or else!"

The official government food illustrations are no-name diets for a reason. Everyone knows that adopting a diet with a name is serious business, signifying a conscious decision to change, to move out of the comfort zone of "normal" eating. The government nutritionists know that placing their recommendations in the category "diet" might be threatening or asking too much of the populace. Instead, they hope to effect change within the confines of the normative diet. But there is an inherent dilemma in trying to get people to adopt something that has no name: there is nothing to hang the take-home message on. Nouns and names help us retrieve things from our inevitably cluttered minds. To overcome this namelessness, the illustrated official diets represent them with a shape, the idea being that a shape can be as effective a symbolic representation of something as a word. The message is that these diets are not really "diets" but simply gentle nudges toward eating a healthy, more balanced assortment of foods every day. But gentle nudges may not work that well, in part because they are pushing against deep cognitive patterns that were laid down in childhood. A less subtle message may be more effective, but then there is the risk of being too demanding or coercive.

The Good and the Bad

In one sense, separating "good" food from "bad" food is pretty easy. Basic criteria include palatability and the ability to sate hunger. If we do not like how a food tastes, or if its taste, smell, or appearance suggests that it might make us sick, we easily classify it as a bad food. Good foods are those whose taste we like or which clearly satisfy an immediate bodily need. As long as it does not make us sick, we can be convinced to eat a food that we do not particularly like. Almost any food can be a good food, because as the old parental admonition goes, if you are hungry enough, you will eat it. But keep in mind that the context in which a food is eaten has a strong influence on whether we think it is good or not.

Good and bad also have moral connotations, and it is here that things get more interesting when it comes to food. Obviously, violating a religiously based dietary prohibition can be seen to be a moral transgression. But at a more subtle level, the foods identified as being bad by nutritional science and public health medicine introduce a moral quandary that is probably unprecedented in human history. Foods that taste good, cause no ill effects in the short term, and are readily available are now being identified as unhealthful and even deadly in the long run. To consume these foods, then, may be seen as a demonstrating a decided lack of willpower, as suggesting a vulnerability to other immoral influences, or even as a form of protracted suicide, another moral violation in many belief systems.

As Michael Pollan has pointed out, it is not foods per se that nutritional science condemns, but substances within the foods.[30] Certain fats, certain carbohydrates, and cholesterol are transformed from being basic constituents of foods to being contaminants. The actual science behind these claims is beyond the scope of this book (see Gary Taubes's impassioned writings for one perspective on that), but there can be little dispute of the claim that official policy (medical, nutritional, and governmental) has led to the vilification of certain foods and substances.[31] Of course, efforts to do this have been countered by those in the food industry with a vested interest in continuing to market foods containing these substances.[32]

Cholesterol stands at the forefront of the vilified food substances, at least in the United States, where official recommendations have prompted a "maelstrom of activity" in clinical, public health, and nutritional circles since the 1960s.[33] Elevated blood levels of total cholesterol and LDL cholesterol are associated with increased risk for cardiovascular disease; consumption of various fats influences blood levels of LDL and HDL cholesterol. Dietary cholesterol has a more muted effect on total cholesterol, especially compared to consumption of saturated fat, and so in many countries health agencies do not recommend a limit on daily cholesterol intake.[34] Yet the United States continues to recommend that daily consumption of cholesterol not exceed 300 mg.

U.S. guidelines for cholesterol consumption were discussed by experts in the field at a 2008 conference (partially funded by the egg industry).[35] The difficulties in implementing public policy based on scientific research were highlighted by some of the attendees. "It was the opinion of some Cholesterol Conference participants that recommendations based on attempts to isolate the effects of dietary cholesterol from complex dietary patterns and extrapolations from complex models are flawed and may lead to unintended negative consequences . . . dietary recommendations stated in scientific language are potentially confusing to the general US population."[36] The American dietary and public health establishment has been very successful at conveying the idea that eating cholesterol is bad: during the 1970s and 1980s, public awareness about cholesterol and heart disease increased substantially and consumption of fatty meats decreased.[37] Perhaps this task was made easier by the fact that cholesterol did not have much of a profile of any kind before it entered the public consciousness as a problematic dietary substance.

One food that became inextricably linked with cholesterol is eggs. U.S. Department of Agriculture data tracking per capita egg consumption in the United States since 1909 shows various ups and downs that can be probably be linked to several factors. The peak of egg consumption was in the immediate post–World War II period (although the uptick in consumption actually began during the war). There was a drop-off in consumption in the 1960s; some of the negative information about cholesterol was beginning to be more widely known during the sixties, and there were other societal changes that no doubt influenced egg consumption (for example, the baby boomers were growing into adulthood and moving out of the house). The 1970s is when many of the anti-cholesterol sentiments in the health community were converted into public policy, and that is when a steady erosion in egg consumption began; in 1975, egg consumption reached the lowest per capita rate since records began being kept in 1909, beating out record low consumption years during World War I. The consumption rate continued to fall into the 1990s, when it bottomed out at around 230 eggs per person per year; this compares to around 400 per year after World War II. Egg consumption has increased slightly since the late 1990s, perhaps reflecting a softening of the anti-cholesterol stance, supported by data showing only a weak link between dietary cholesterol and levels of total cholesterol in the body.

Whatever the pros or cons of limiting cholesterol intake, it is clear that cholesterol was successfully demonized and foods associated with it were (and are) considered to be "bad," at least in the minds of the segment of the population that pays moderate attention to dietary health. It remains gospel that dietary cholesterol is to be avoided.[38] This is all quite remarkable given that cholesterol is frequently associated with foods that are perceived to taste good. Many people clearly developed an aversion to foods containing cholesterol, or tried to, hence the obvious success of marketing cholesterol-free versions of foods or even labeling foods as such even though they clearly did not have any cholesterol in them to begin with (such as all non-animal products). Ironically, foods high in saturated fats were somehow sanctified by a "cholesterol-free" label.

An aversion to a food can lead to that food being considering disgusting. Foods one does not eat also can become disgusting (which Marvin Harris used as the basis of his argument about why Westerners feel disgust when it comes to eating insects). Disgust has strong evolutionary roots: foods that promote vomiting or nausea are dangerous and are regarded as disgusting upon subsequent encounters. But in humans, basic cognitive processes are almost always elaborated by culture. In a cross-cultural analysis of disgust, Jonathan Haidt and his colleagues argue that the basic feeling of disgust can exert a powerful influence on cultural behavior and even cultural institutions:

> We have argued that core disgust is an emotion that makes people cautious about foods and animal contaminants of foods. We have argued that disgust has extended among Americans to become not just a guardian of the mouth, but also a guardian of the "temple" of the body, and beyond that, a guardian of human dignity in the social order. And

finally, we have argued that this expansion, from food to the social order, is not unique to Americans, but can be found in some form in many cultures.[39]

This is another instance of the two-way street between culture and cognition: the basic brain wiring of evolution creates culture but in turn is remapped under cultural influences.

The various attendant bodily sensations of disgust can function as what Antonio Damasio calls a "somatic marker," also commonly called a "gut feeling."[40] As Damasio writes, "When the bad outcome associated with a given response option comes to mind, however fleetingly, you experience an unpleasant gut feeling."[41] Somatic markers help us make decisions because they alert us quickly to the possible positive or negative outcomes to a contemplated course of action. The feelings associated with disgust are a very powerful somatic marker. They virtually compel a course of action: avoidance.

Damasio's somatic marker hypothesis highlights the intimate connection between mind and body. The sensations of the body, experienced during feelings and emotions, play a prominent, if sometimes covert, role in all manner of conscious cognitive processes, including those important in decision making and social behavior. Body and culture are thus linked, not only by traditional foodways that develop in every culture, but also in a dynamic and functional way by the mind itself. Moral decisions about what is good and bad—about what *feels* good and bad—are made with somatic input. Food and eating prompt profound feelings and changes in the body; it is not surprising that it is quite easy to forge a link between food and morality.

The functional brain anatomy of morality—linking these very specific kinds of feelings to the brain—has drawn increasing interest from cognitive science researchers.[42] As one might expect, depending on the moral phenomenon in question, different networks in the brain may be activated. One focus of study has been on the relationship between disgust and morality. In a fascinating study, Thalia Wheatley and Jonathan Haidt tested the somatic marker hypothesis by hypnotizing subjects and introducing a posthypnotic suggestion that would cause them to feel "a brief pang of disgust . . . a sickening feeling in your stomach" upon reading a particular word.[43] Half of the sixty-four subjects were induced to feel disgust upon reading the word *take*, the other half for *often*. Each subject read six vignettes concerning some moral situation, half of them with their "disgust" word inserted and half without. After reading the vignettes, the subjects were asked to rate each one for the level of disgustingness they thought the scenarios expressed. The results were quite strong: if the posthypnotic disgust word was present in the vignette, the subjects were more likely to give it a higher disgust score. When subjects were asked to make comments on their ratings, they said things such as "I knew about 'the word' but it still disgusted me anyway and affected my ratings. . . . It just seems so weird and disgusting. . . . I don't know [why it's wrong], it just is." Subjects were also asked to rate the vignettes on a morality scale; although the results were not as strong as for disgust, when the disgust word was present, subjects were more likely to rate the vignette as depicting something morally wrong.

Wheatley and Haidt argue that their results show strong support for the somatic marker hypothesis; in this case, the artificially induced feeling of disgust not only affected whether something was seen as disgusting but also colored its moral perception. Does functional neuroimaging support a link between disgust and morality? A study by Jana Borg and colleagues indicates that the answer is yes and no.[44] They looked at regions of the brain activated when subjects were presented with disgusting stimuli (verbal statements presented in the context of a standard memory-recall test) centered on pathogens (eating unclean or contaminated substances), incest, and non-sexual moral situations, which were compared with each other and to a neutral situation. The results clearly showed that disgust relating to pathogens and to sociomoral acts involve overlapping brain networks. These networks were quite widespread, including parts of the basal ganglia, the amygdala, and several cortical regions. However, the pathogen-related stimuli elicited activation in regions not activated by the other situations. In addition, processing the incest and non-sexual scenarios activated both overlapping and distinct regions.

Again, it is not surprising that different moral situations are at least partially processed in different networks in the brain. After all, a moral situation is a very complex stimulus that calls on a variety of cognitive domains, including those involved with the senses, memory, emotion, and so on. From the perspective of food preferences and morality, these results show why it is easy for a food to go from being a bad or good food in terms of palatability or healthfulness to being an exemplar or embodiment of some aspect of a culture's moral systems. This does not always happen—sometimes broccoli is just broccoli. But clearly, morality and disgust interact at many different cognitive levels. Foods that are morally repugnant, such as dog or rat to a Westerner, become disgusting without ever being tasted and rejected. Substances such as cholesterol can become fixed as being bad and to be avoided even if they do not really have any taste profile at all.

People are often disappointed that official recommendations about what foods are healthful or not seem to change in a cyclical fashion. First all fat is bad; then some fat is good. We hear that cholesterol should be avoided; then we're told that dietary cholesterol doesn't matter. Dietary recommendations are easily conflated with moral pronouncements, a result of co-opting the cognitive machinery of disgust in the brain for non-food-related purposes. Food choice thus often has a moral dimension, sometimes explicit but usually implicit. So when dietary recommendations are changed, exposing the shaky basis on which they were originally formulated, it can be not just annoying but potentially distressing.

Menus in Your Mind

In the 1950s, food writer Joseph Wechsberg described a Viennese restaurant called Meissl & Schadn, which was one of the many casualties of World War II.[45] Wechsberg suggests that it started dying long before then, however, its style and menu a relic of the days of the Habsburg Empire; it managed to survive through the 1920s on momentum and the force of will of some of its employees. Meissl & Schadn was known worldwide for its boiled beef, of which it served no less than twenty-four varieties: *Tafelspitz, Tafeldeckel, Rieddeckel, Beinfleisch, Rippenfleisch, Kavalierspitz, Kruspelspitz, Hieferschwanzl, Schulerschanzl, Mageres Meisel* (or *Mäuserl*), *Fettes Meisel, Zwerchried, Mittleres Kügerl, Dünnes Kügerl, Dickes Kügerl, Bröselfleisch, Ausgelöstes, Brustkern, Brustfleisch, Weisses Schezl, Schwarzes, Scherzl, Zapfen,* and *Ortschwanzl*. This was a menu for experts, both "concise and ambiguous at the same time." As Wechsberg points out, you do not go to Tiffany's and ask for a "stone"; the patrons of Meissl & Schadn did not go there and ask for "boiled beef." The menu reflected the minds of its patrons and their intimate knowledge of the butchered anatomy of the steer.

I am not sure if I know twenty-four types of any kind of food. In my youth, I worked at a Baskin-Robbins 31 Flavors ice cream

store, so I am aware that at one time I did know quite a large
number of ice cream flavors (many more than thirty-one because
of the seasonal rotation of flavors). I am sure that any of us working there at that time could have produced a reasonable taxonomy
of flavors: chocolate-based, vanilla-based, coffee-based, sherbets
and ices, seasonal and other oddities, caramel or butterscotch
swirls, chocolate chips. As teenage experts in ice cream flavors,
we would have disagreed with one another on certain points, but
to some extent the expert's quest is to push classification to its
limits and expose ambiguities and subtle distinctions. I am quite
certain that all of us would have agreed that bubble-gum-flavored
ice cream was the aardvark of our world: a species classified only
with itself.

Almost all menus, with their various categories of food, reflect
both explicit and implicit classification systems. This becomes
obvious when we are confronted with twenty-four kinds of boiled
beef, thirty-one flavors of ice cream, or fifty kinds of sushi. These
extreme menus, with their overwhelming representation of types
from a single category, poignantly illustrate the human penchant
for classification. Whether or not we wish to engage in such an
exercise, it is hard not to be impressed by the identification of
so much variety. The ability to *deliver* all that variety is another
matter. We expect that an elaborate level of classification should
be matched by an equally expert level of preparation; when it is
not, we tend to look at the menu as promising something more
than it could deliver.

From the simple menu that Escoffier made from his battlefield meal to the celebration of boiled beef featured at Meissl &
Schadn, all menus are external reflections of how food is organized in a chef's mind, or in the case of corporate restaurants,
in the collective mind of the committee of chefs, food scientists,

and marketing consultants that worked to make a menu. Menus are formalized maps of how brains organize food. They are formalized by culture, which expects certain items to be listed in certain ways, no matter what the idiosyncratic bent of the menu maker.

The menus in our minds have more freedom. Yes, we might have internal food categories that match what is generally out there in our culture: meat, vegetables, appetizers, mains, desserts, and so on. But we should also have other categories in our own personal menus: foods I like on my birthday, things I will eat as leftovers and things I won't, main courses for important guests, foods that make my children feel better when they are sad, foods my dog won't eat, foods I secretly love but am ashamed to admit publicly to liking, foods my wife grew up on, and so on. Food is not unique in its ability to stimulate distinct and overlapping categories in our minds. Indeed, classification and categorization are essential to having a mind that works correctly. But food is ubiquitous and evocative at all of the different levels of human cognition. It is a worthy exercise in self-knowledge to consider how we think about food categories, what they mean, and how they change and become more or less important over the course of a lifetime. Our internal menus may not feature twenty-four kinds of boiled beef, but it may be surprising to find that they feature an even more extensive classification of items reflecting all aspects of the human experience.

7 ⟋ᴓ⟍

FOOD AND THE CREATIVE JOURNEY

IN THE 1950s, Japanese researchers started to provide sweet potatoes and other foods to a troop of macaque monkeys living on a
small island off the coast of Kyushu, the southernmost of Japan's
major islands.[1] They did this initially to habituate the monkeys,
to make them more comfortable with human observers so that
their behavior and habits could be more easily studied. Over time,
however, it became an experiment to examine how the monkeys
transmit knowledge and learn about food. One of the monkeys, a
young female nicknamed Imo (otherwise known as 111), was seen
dipping sweet potatoes into the surf before eating them. This
curious habit made perfect sense: not only did it wash the sand
off the potatoes, but she clearly liked the salty flavor that the seawater added. Soon, except for the oldest members of the troop,
all of the other monkeys followed Imo's lead and started washing
their sweet potatoes in seawater. The researchers then gave the
monkeys kernels of wheat, which they scattered across the sand.
Imo discovered that by scooping up handfuls of the wheat and
sand and tossing them in shallow puddles and pools on the beach,
it was possible to easily separate the wheat from the sand, because
the wheat would float while the sand would sink. Again, this
behavior was adopted by others in the troop.

Imo was a very creative monkey, and her exploits have earned her a kind of immortality in anthropology textbooks. By the time I had the opportunity to visit Kojima, the island where the monkeys lived, in the late 1980s, food provisioning by the researchers had been scaled back considerably. The research on protocultural behavior in the monkeys was over and the island, reachable from the mainland at low tide, was becoming overpopulated with monkeys (a more distant small island, much more difficult to reach, had a population of one very fat monkey). The researchers did on occasion still throw some wheat on the beach, and I got a chance to observe this while I was there. It was exciting to see the monkeys do what they were famous for. Rather than carrying handfuls of the wheat-sand mix to a puddle, most of them used their hand to sweep the wheat off the sand into a nearby puddle. The technique worked fine, as the sand quickly separated from the wheat kernels and the monkeys could pluck the kernels out of the water one by one. I brushed a bit of wheat into some water myself, although I did not eat it. It was a rather unique experience to be following a recipe invented by a monkey and relive a seminal creative moment.

Creativity in food preparation—creativity of all kinds—has come a long way over the course of human evolution. One of world's greatest chefs is Thomas Keller, the mastermind behind two of the most celebrated restaurants in the United States, the French Laundry and Per Se. Keller is widely recognized for his almost unparalleled dedication to the craft of cooking. Having cooked from his recipes and Imo's, I can attest that his tend to be somewhat more elaborate and complicated. Keller has reflected on the nature of the creative experience in cooking: "You're not going to be able to duplicate the dish that I made. You may create something that in composition resembles what I made, but more important—and this is my greatest hope—you're going to create something that you have deep respect and feelings and passions for. And you know what? It's going to be more satisfying than anything I could ever make for you."[2] I appreciate Keller's sentiment here: he knows that he is the creative force behind his recipes, but that does not mean that in re-creating them, we cannot also experience some of the excitement and reward of creativity. In a sense, following his recipes is a creative act in the same way that a classical musician playing a piece is considered to be a creative artist, even if he or she does not write music at all. The sense of creativity is likely enhanced when, in following a recipe or a particular musical score, the creative element is in the foreground: creativity itself can be inspiring. Knowing Imo's story of creativity helped infuse the act of brushing wheat into water with more meaning for me. It was a simple recipe, but then sometimes the most creative acts are the simplest.

Keller's standing among chefs is unchallenged, but he does not carry the responsibility of being the standard-bearer for creativity among the elite chefs of the world. That honor belongs to Ferran Adrià, head chef at El Bulli, the world-famous restaurant (now on hiatus) located outside Barcelona.[3] Adrià is noted for the assortment of foams, essences, oils, and other deconstructed and reconstructed forms of food that he serves along with the most meticulously prepared and presented conventional ingredients. Adrià places creativity at the center of his culinary universe, recognizing that when people come to eat at his restaurant, they expect to have something new and exciting. I could try to describe one of his dishes, but it would be difficult to do it justice (see *A Day at El Bulli* for photographs of dishes and a detailed exploration of the entire production process at the restaurant).

Adrià has written extensively about his own pathways to and ideas about creativity. Here are a few of his pithy aphorisms:

224 THE OMNIVOROUS MIND

New, creative, and unique are not the same thing.
With creativity, it is not what you look for that matters,
    but what you find.
Creativity means changing your mind every day.
To be truly creative, a dish must be interesting as well as new.

Adrià sees creativity as a multistage process, which forms the basis of the El Bulli approach to food and cooking. This approach begins with a strong foundation in traditional cuisines, both local and distant, and a thorough mastery of core techniques applied to a variety of foods. It gives equal weight both to the search for new ingredients and to using old ingredients (including commercial foods and additives) in new ways. Dishes and menus are reconfigured and remixed. There is an appeal to all of the diner's senses, including a "sixth sense"—a direct connection to the diner's mind to evoke something from his or her memory or experience.

At El Bulli, the engine of creativity is fueled by a quintet of orienting principles: association, inspiration, adaptation, deconstruction, and minimalism.[4] Association is the most basic of these principles, and refers simply to combining ingredients in new and perhaps unexpected ways. Inspiration is the vaguest of them. Adrià says that inspiration for a dish can come from virtually anything—from a flower peeking through grass to a modern painting. Adaptation refers to making something old new again by modifying its presentation, substituting ingredients, or even altering its flavor profile from sweet to savory and vice versa. Deconstruction can be a special form of adaptation, in which the basic elements of a dish are present but rearranged or dispersed while still making reference to an original template; it can involve taking a set of ingredients used to make one dish and transforming them into another dish. Minimalism is, as might be expected, the challenge to make a powerful or engaging dish with the fewest number of ingredients possible.

The Japanese macaque Imo was probably limited by her native intellect to fairly simple recipes. After millions of years of cognitive evolution, we humans have arrived at a place where culinary minimalism is a creative choice rather than a necessity (at least in the rarefied kitchens of Michelin three-star restaurants). Humans are a creative species. When we compare what we do and make with what other species do and make, it is self-evident that we are the most creative species. But what role did creativity, in relation to food and other aspects of our ancestors' lives, have in human evolution? Is it just a by-product, an emergent property, of more fundamental cognitive processes, such as those involving problem-solving ability, memory, language, and attention? Or is creativity a distinct and isolable cognitive process of its own, one that varies from individual to individual or which perhaps can be nurtured in the proper environment?

Creativity in Evolution: What Is It Good For?

Grant Achatz is one of America's great chefs. Even before it opened in 2005, his restaurant in Chicago, Alinea, was already being hailed as a culinary destination on par with Thomas Keller's and Ferran Adrià's restaurants. At the time of the opening, Achatz was the thirty-one-year-old wunderkind of American cooking, a disciple of both Keller (by virtue of having worked in one of his restaurants for four years) and Adrià (as a source of inspiration). In its reliance on technology and chemistry as vehicles for culinary creativity, Achatz's cooking appears to owe more to Adrià than to Keller.[5] However, Achatz's cookbook, *Alinea*, makes it clear that it is perhaps better to look at his cooking as a technological expansion of the Keller approach. According to critics, there can be no doubt of Achatz's astonishing creativity and culinary skill. His restaurant has received the highest accolades, and like Keller and Adrià, Achatz has produced an oversized, photo-stuffed cookbook featuring his dishes, his philosophy of cooking, and his thoughts on the creative process.

In terms of his creativity, it is clear that Achatz need not take a backseat to any other chef. Yet it is also clear that he had his antecedents, and that his individual creativity has been strongly influenced by the works of others. Achatz himself acknowledges that in cooking, as in many other fields, change comes gradually, but once in a while a significant individual—an Escoffier or an Adrià, for example—prompts a paradigm shift. He sees himself as the beneficiary of Adrià's creative revolution, which in turn has spurred him and other chefs, each with their own unique backgrounds, to greater heights.

We all benefit from the artistic creativity of others. We enjoy eating the food of creative chefs, listening to the music of great composers and songwriters, reading the books of perceptive writers, looking at the paintings and sculptures of imaginative visual artists. These types of works were for centuries mainly enjoyed only by elites in large-scale civilizations, but the modern media culture makes creative products available on a historically unprecedented scale. In contrast, over much of the course of human evolution, creativity could be beneficial only at a face-to-face level. In this context, perhaps the most significant contribution of creative individuals over evolutionary time has been when they show us all how to do something in a new way; when this occurs, creativity is not simply passively enjoyed but becomes an integral part of an individual's life. The impact of the Japanese macaque Imo's creativity was felt not when the other monkeys passively observed her washing sweet potatoes or floating wheat but when they followed her lead and copied her actions. Innovation was propagated by social learning.

Herein lies a pretty basic evolutionary dilemma. Primates, including monkeys, apes, and especially humans, are great at learning from the actions of others. In a natural or traditional setting, when an individual creates something, it is out there to be freely copied by others. Thus what we typically think of as creativity would have little chance of conferring increased reproductive fitness to a creative individual, and direct natural selection for creativity would have been very difficult over the course of human evolution. As Gregory Cochran and Henry Harpending summarize the situation: "Creativity seldom confers large fitness advantages, because good new ideas can be rapidly copied by others. The copiers receive the fitness benefits without paying the associated costs."[6] The current social environment, in which creative ideas and the associated economic benefits derived in the marketplace are protected by law, is very unusual and very recent.

But maybe we do not have to be so negative about the role of creativity in human evolution. Certainly, at the most exalted levels of creative genius, the opportunities for selection would be few and far between. Creativity researcher Dean Keith Simonton outlines just how rare creative genius would have been in the relatively small population groups that characterized most of human evolution.[7] He argues that these small groups would have trouble satisfying the several prerequisites, both biological and social, necessary for the full expression of the highest levels of creativity: a pool of individuals with the required combination of intelligence and creativity; opportunities for all potential creative geniuses to fulfill their promise (e.g., the potential pool decreases if entire groups, such as women or slaves, are socially excluded);

social roles that allow or encourage creative expression; and education to provide the opportunity (the building blocks of knowledge, previous discoveries, etc.) for creative individuals to learn how to be creative. Simonton argues that the efflorescence of creativity that we associate with the emergence of large-scale civilizations compared to traditional, small-scale societies is due largely to population growth. Larger populations generate a greater number of potentially highly creative individuals as well as the roles and opportunities that encourage creative expression.

Creativity is not always about creative genius or artistic creativity, however. Creativity can be recognized at a much more mundane level. Alice Flaherty provides a definition that reflects how creativity is investigated in neurological and psychological research: "A creative idea will be defined simply as one that is both novel and useful (or influential) in a particular social setting."[8] This is more the Imo level of creativity than the El Bulli level, but it is the kind of creativity that would have been more important over the long haul of evolution. Evolutionary psychologist Geoffrey Miller argues that everyday creativity may pay off in smaller social settings and in sexual competition.[9] Humor and other forms of entertainment may be examples of this more personal, yet potentially reproductively advantageous, creativity.

Monkeys and apes are capable of coming up with novel and useful solutions to the technological and social problems they may face. Humans, however, have taken this more basic level of primate competence to a much higher level. One factor that gives humans a creative advantage is language. The innovations of a chimpanzee or macaque can be communicated only through direct observation. These innovations can be preserved by being stored in the minds of individuals, who can passively transmit them again only if they are directly observed by other animals while engaged in the creative act. Although this is effective up to a point, direct observation imposes serious constraints on the kind, number, and complexity of innovations that can become part of the group's common knowledge.

With language, humans developed a means by which the products of creative thinking and creative ideas could be communicated and preserved. At some point during our evolution, language-based cultures provided our ancestors with a vehicle for storing and sharing innovative ideas. Some of these ideas, such as how to process fallback foods that could be eaten during, say, a prolonged drought, might be useful only once in a generation, but when used, they may be of critical importance to the survival of many individuals. Language allows creative ideas to be carried forward in time and space. It completely changes the environment in which creativity is expressed.

Imo's troopmates learned to wash sweet potatoes and wheat kernels by observing her; this is a form of imitative social learning. However, Susan Blackmore and others researchers argue that there are some fundamental differences between the macaques' behavior in this setting and the more complex forms of human imitation.[10] Imo's troopmates were copying not simply her behavior but her interactions with specific objects in their environment. Humans do the same thing, but they can isolate the behavior from the object and context: a human would have no trouble rehearsing the washing behavior even if he or she did not have a sweet potato and was away from the beach. This would be beyond a macaque's ability, just as chimpanzees would probably find it nonsensical to rehearse termiting behavior if they were not holding a stick in proximity to a termite nest. Humans can easily separate the behavior from the object and setting.

Language, the ability to imitate complex multistep tasks, an increased cognitive storage capacity—all build on a basic primate ability to innovate, which in turn confers on humans a uniquely enhanced level of creativity. Yet Cochran and Harpending's contention that selection for individual creativity is problematic still holds. Groups benefit from the creativity of their individuals, which in turn undermines the fitness benefits that an individual within the group may accrue from his or her creativity. This suggests that selection for creativity may be a function of selection at the group level more than at the individual level, a controversial topic in evolutionary biology. But groups cannot have too high a proportion of creative individuals, in that part of what makes cultural groups effective is their traditional conservatism.

We see that, on several levels, creativity exists in the balance. The idea that there is a thin line between madness and genius is just an extreme example of this balancing act. Cultures must find a balance between indulging creative expression and maintaining traditional practices. Individuals must find a balance between knowing when a creative solution to a problem is warranted or necessary and when the tried and true works best. Human societies may work best with a balance of creative and non-creative individuals. It is clear that people vary according to their creative inclinations. How these inclinations are developed and fulfilled is another matter, depending on the creative environment as well as some native tendency toward creativity.

What can archaeology tell us about the evolution of human creativity?[11] The archaeological record from the past 10,000 years, since the advent of agriculture, is rich with the ruins and remains of great and near-great civilizations, as well as other strong signs of the human propensity to leave a mark on the landscape. But beyond this, dating back to at least 100,000 years (corresponding to the Late Stone Age in Africa and the Upper Paleolithic in Europe), we see the material evidence of a "creative explosion," as it has been called, rendered in stone, in bone, and with paint on the walls of caves and cliffs. The array of new forms of tools, along with symbolic and representational art, speaks clearly to the arrival of fully modern humans capable of producing a rich material culture.

Going further back in time, the archaeological record offers considerably less evidence of creativity. The human and chimpanzee lineages split from each other about 6 million years ago. The hominin archaeological record begins about 2.5 million years ago with very crude stone tools. These tools get better over time, becoming more sophisticated, but the classic handaxe-using form of *Homo erectus* remained remarkably stable over a very long period of time. This does not mean that creativity, intellect, or whatever else it is that we might be looking for did not change over this time period. It is just that there is little evidence for it in the artifacts that we find from that time.

Archaeology isn't just about what is found; where it is found is important as well. Perhaps the earliest evidence of exceptional hominin creativity, above and beyond that seen in our great ape cousins, can be seen starting about 2 million years ago. It was at about this time that the earliest members of the genus *Homo* begin their expansion out of Africa and into other parts of the Old World. This expansion was quite rapid, and by 1.8 million years ago *Homo* species were found across a wide expanse of Eurasia. If we keep in mind that the current great apes live in or near tropical forests, the spread of our ancestors out of these environments, unaccompanied by any similar ape migration, suggests that early *Homo* was more effective at coming up with solutions to the problems posed by novel or changing environments. The most critical

problem facing any animal moving to new environments is finding food. Sometimes this sort of expansion is a product of following the food (e.g., hunters following prey as they migrate). This
may have been true of early *Homo* as they developed a diet more
reliant on animal food than the diets of the great apes, but it
is unlikely to be the whole picture. More generally, I think this
migration was indicative of the first steps toward the superomnivory that characterizes our species (see Chapter 2). This superomnivory reflects our ancestors' ability to more creatively exploit
the resources of varied environments, adapting behaviorally rather
than physiologically to new foods.

Creativity and the Brain

A bottom-up appraisal of creativity begins with the brain. Understanding not just the evolution of creativity but also how and why
some people are more creative and others less so requires an investigation of the brain structures and networks underlying creativity. This may seem unduly reductionist to some (you know who
you are), but understanding the neural basis of creativity should
not make us appreciate it less. Furthermore, it is important to keep
in mind that given our understanding of brain plasticity, there is
no reason to think that creative neural processes are exclusively
hardwired; they may be nurtured with a proper environment. Undoubtedly there are individuals who are born with the whole cognitive package required to be exceptionally creative, but creativity
is not just about creative genius. Problem solving is something that
the human species is good at; the ability to come up with novel
and effective solutions to problems is always a good thing.

Psychologists have studied creativity for decades. One of the
basic models for the creative process was proposed in 1926 by

Graham Wallas.[12] Like much of the psychological science of that time, Wallas's model was based more on introspection than experimentation, but its influence and durability suggests that he was on to something. Wallas proposed a four-stage model of the creative process. The first stage is *preparation*, in which the problem is identified and set up, and the creative individual consciously brings his or her expertise and problem-solving ability to the issue. The second stage is *incubation*, during which the mind is not consciously working on the problem but subconsciously forms and assesses associations that may be relevant to solving the problem (or, in the case of aesthetic creativity, selecting among good or bad ideas that may be further developed). The third stage is called *illumination*, the aha moment when a creative solution moves from the subconscious to the conscious. Finally, there is *verification*, during which the incipient idea that emerges during illumination is consciously assessed, refined, and developed. This process is not linear but potentially recursive, with earlier stages being revisited as necessary during the process of creativity.

The four-stage model of the creative process has been useful, and for many years it represented the primary launching point for research in the psychology of creativity. But as Todd Lubart points out, no model of the creative process satisfactorily accounts for all kinds of creativity.[13] Of course, there should be no expectation that a one-size-fits-all model of creativity is adequate to describe what goes on in our brains. From an evolutionary perspective, if creativity has any value, then it is as a manifestation of the behavioral flexibility of our species. Behavioral flexibility does not imply dependence on a single, rigid cognitive process, but rather suggests the ability to combine cognitive processes in creative ways to deal with dynamic or novel social and ecological environments. The complex psychology of creativity almost ensures that there is no single "creative center" in the brain.

Insights about the brain and creativity have come from a variety of sources, including the study of brain diseases, brain lesions, and modern neuroimaging. One disease, called frontotemporal lobar degeneration (FLD), has been of particular interest to neurologists studying the neurobiology of creativity. This condition is just as it sounds: a progressive atrophy of the frontal and temporal lobes that can lead to dementia and other cognitive problems. Any disease that affects the frontal lobes would be expected to cause a deficit in creativity. Yet some researchers have reported clinical cases in which artistic talent and creativity increased with the onset of FLD.[14] Individuals who had no interest in art before the onset of illness became painters afterward. In a patient who was already an artist, technique improved with the disease, although the ability to finish a painting decreased. Other researchers argue that FLD patients are not "productively" creative.[15] Their performance on standardized psychological tests of creativity tends to be poor, which reflects their inability to constructively organize their thinking (i.e., a frontal lobe deficit).

The "pseudo-creative" responses of FLD patients may not be evidence of emerging talent or creativity, but rather may be a result of disinhibition caused by frontal lobe damage. For example, the use of sexual imagery or perseveration in FLD patients may be assessed subjectively to be an indication of increased creativity, but these are not necessarily indicative of truly creative work.[16] On the other hand, increased disinhibition may be an important personality component of creativity. Part of what makes a creative person creative is a willingness to challenge the accepted way of doing things, to have a certain fearlessness in the face of those who would not change the status quo.

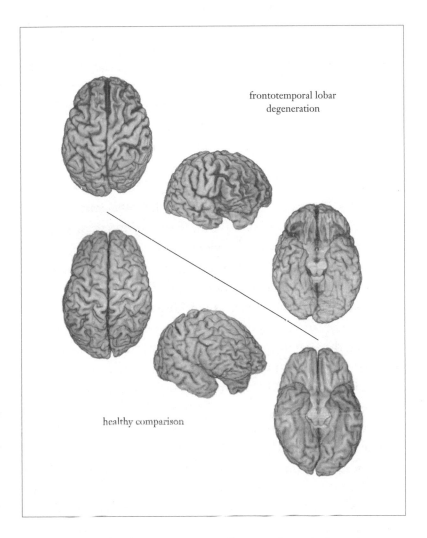

frontotemporal lobar
degeneration

healthy comparison

Frontotemporal lobar degeneration eventually causes dementia due to the wasting away of the frontal and temporal parts of the brain. Note the widening of the sulci in the FLD patient compared to the healthy comparison subject. In milder cases, disinhibition in FLD individuals may foster creative expression.

Neurologist Alice Flaherty has proposed a model of creativity that takes into account anatomical information from neuroimaging and lesion research, combined with a neurochemical perspective provided by study of pharmacological effects on creative thinking.[17] Flaherty's model attempts to explain not only artistic creativity, such as painting or music, but also creativity in other domains, such as language, science, and mathematics. Expanding on some of the research on FLD patients, Flaherty suggests that a network linking the frontal and temporal lobes is indeed especially important for creative processing. More specifically, interaction between regions within these two lobes appears to act as a regulator of creative expression. Temporal lobe deficits can increase the generation of creative ideas, sometimes at the expense of quality, as occurs when someone is in a manic state. In contrast, frontal lobe deficits can inhibit creative thinking, since the prefrontal cortex is essential for many of the cognitive processes necessary for creativity, such as working memory, sustained attention, abstraction, and planning. Thus via mutually inhibitory pathways, the frontal and temporal lobes work together to generate ideas that are both novel and useful.

In addition to that frontotemporal network, Flaherty emphasizes another brain system critical for creative expression: the dopamine pathways of the limbic system (which have strong connections to the frontal lobe). Being creative is not a passive process. Creative people are more responsive to sensory stimulation, have higher baseline levels of arousal, and display more intense goal-directed behavior. In Flaherty's model, one of the factors underlying variability in creativity can be traced to the levels of activity in the dopamine system. As we have already discussed, dopamine mediates reward-seeking behavior, including the sense of appreciation for aesthetically pleasing stimuli such as beautiful faces and music. Flaherty suggests that creative motivation is also strongly influenced by these dopamine pathways. The evidence for this comes primarily from pharmacological research. Drugs that increase the activity of dopamine (agonists), such as cocaine and levodopa, heighten arousal and goal-seeking behaviors, while dopamine antagonists, such as some antipsychotic medications, can shut down the free associations that may be necessary for creativity.

Motivation and an internal monitor of the idea generation process are clearly essential parts of the neurology of creativity. Studies of brain anatomy and creativity indicate, however, that other regions of the brain show changes that correlate with higher or lower levels of creativity. Psychologist Rex Jung and his colleagues looked at how the thickness of the cortex in various parts of the brain correlated to performance on a psychological instrument called the Creative Achievement Questionnaire.[18] They found several brain regions in which cortical thickness showed statistically significant correlations with performance on these tests of creativity and divergent thinking. These regions were distributed in the cortex of both hemispheres and included parts of the frontal lobe and regions on the border between the temporal and occipital/parietal lobes. The results of these studies strongly support the hypothesis that creativity in the brain depends not on a specific region or hemisphere but on a dispersed network of brain regions. Interestingly, there were some brain regions where cortical thickness was negatively associated with creativity scores.

Other anatomical studies support the idea that dopamine networks are indeed important in creative thinking. Strong associations have been found between performance on a creativity test and gray matter volume in dopamine-rich subcortical regions such as the substantia nigra and other frontostriatal regions; in addition, a portion of the right dorsolateral prefrontal cortex was also correlated with higher creative test performance.[19] In the thalamus, the important subcortical gateway to the cortex, it has been found that lower density of a certain kind of dopamine receptor is associated with increased creativity (more specifically, on a test of divergent thinking).[20] Örjan de Manzano and colleagues suggest that the reduced activity of dopamine in this region leads to a reduction of inhibition and enhances the flow of information between cortical regions. In healthy individuals, this could form the physiological basis for enhanced creativity. The risk here is that these receptors are also implicated in the development of psychopathology. So, for example, if the inhibitory function of the thalamus is too impaired, cortical areas can become overwhelmed by excessive excitatory signals, leading to dysfunctionally disinhibited behavior.

In terms of brain anatomy and creativity, one of the most pervasive ideas in popular thinking is that the right hemisphere of the brain is the creative side, while the left is more analytical. As Alice Flaherty points out, this idea is biased at the outset toward certain kinds of creativity, such as the visual arts.[21] It is also a notion that has been around for a very long time, originally emerging in the nineteenth century, when medical researchers developed a whole series of polarities concerning the hemispheres.[22] These polarities reflect ancient themes in Western thought that see human existence as an uneasy compromise between opposites. According to historian Anne Harrington, as nineteenth-century neurologists began to explore the brain and behavior, they imported this notion of behavioral duality to interpret the obvious anatomical duality of the brain. The dominance of the left hemisphere for both language and handedness was extrapolated to other domains. Thus the left side was seen as the seat of reason and consciousness, while the right was burdened with madness and

unconsciousness. Or the left was maleness, intelligence, and voli-
tion, while the right reflected femaleness, passion/emotion, and
instinct. It was quite obvious to these nineteenth-century medical men that the left side was the "good" or human side in most of
these polarities.

The pervasive idea of the unique capabilities of the left and
right hemispheres was easy enough to apply to creativity, especially following the famous split-brain studies that Roger Sperry
conducted in the 1950s and 1960s. Sperry and his colleagues'
groundbreaking research on patients who had had most of the
connections between the two hemispheres surgically severed (as
a treatment for intractable epilepsy) showed for the first time
under controlled conditions just what each hemisphere was up to.
In his Nobel Prize lecture, Sperry summarized his group's discoveries about the right hemisphere:

> The right hemisphere specialties were all, of course, non-
> verbal, nonmathematical and nonsequential in nature. They
> were largely spatial and imagistic, of the kind where a sin-
> gle picture or mental image is worth a thousand words.
> Examples include reading faces, fitting designs into larger
> matrices, judging whole circle size from a small arc, dis-
> crimination and recall of nondescript shapes, making men-
> tal spatial transformations, discriminating musical chords,
> sorting block sizes and shapes into categories, perceiv-
> ing wholes from a collection of parts, and the intuitive
> perception and apprehension of geometrical principles. The
> emphasis meantime became shifted somewhat from that
> of an intrinsic antagonism and mutual incompatibility of
> left and right processing to that of a mutual and supportive
> complementarity.[23]

Note that Sperry was careful to emphasize the complementarity between the two hemispheres. Many people who read Sperry's work immediately grasped its significance as showing that the right hemisphere's capabilities reflected the more artistic, especially visual, aspect of the human condition, which they presented in opposition to the analytical rationality of the left hemisphere. At the popular level, there was a tendency to interpret Sperry's research in nineteenth-century terms.

So is there any neuroimaging evidence for a right-side bias in creativity? The structural data appear to provide little support for it, but functional studies suggest some rightward leanings in the creative process. Mark Jung-Beeman and his colleagues conducted a functional MRI study in which subjects were asked not only to solve a word problem but also to indicate whether they solved the problem with or without insight.[24] In this situation, insight corresponds to the aha or eureka moment that people have when they realize that they have come up with a solution to a problem seemingly out of the blue. Jung-Beeman and colleagues found that when subjects said that they had an insightful solution to a problem, a region in the right temporal lobe showed increased activation.

This part of the right temporal lobe is important for semantic integration, associating information (especially verbal) from conceptually distinct domains. Both the left and right temporal lobes do this sort of thing, but befitting its role as the "language hemisphere," the left side does it in a much finer and more focused manner. Semantic integration in the right temporal is much coarser and weaker, and may involve neural processing fields that are much less well-defined and more overlapping than in the left hemisphere. It is for just these qualities that Jung-Beeman and colleagues hypothesize that the right temporal lobe is associated with the aha moment. The combination of weaker and coarser semantic integration means that processing occurs below the conscious level, but in terms of creativity, the overlapping processing fields may allow for the creation of associations more readily than in the left temporal lobe. These subconscious associations are directed toward solving the problem, but the actual processing is not perceived until a solution is arrived at in the eureka moment.

This study by Jung-Beeman and colleagues illustrates a rightward creativity asymmetry on a very specific task in a very localized part of the brain. The authors in no way claim that this means that the right side of the brain is the creative side. In fact, the circumstances of the task in this experiment highlight the complex relationship between brain processing and creativity: a region in the right temporal lobe showed increased activation for a language-related task, which is generally considered to be more the domain of the left hemisphere. Other studies show evidence for bilateral and asymmetrical activation for creative tasks; some suggest that people who are more creative have a greater integration of the two hemispheres rather than an increased reliance on one or the other.[25] A statistical review of a large number of functional brain studies does suggest there may be a rightward bias for creative thinking.[26] This bias may be related to greater right hemisphere activity during abstract thought. Of course, this does not mean that creativity is solely the function of the right hemisphere, or that people who are more creative are more dependent on the right hemisphere than those who are less creative.

One last aspect of brain anatomy that may vary with creative ability is the white matter, or at least certain pathways within the white matter. Rex Jung and his colleagues have used diffusion tensor imaging, a modification of MRI that allows for the assessment of white matter integrity, to look at the relationship between creativity and the pathways through which neurons communicate.[27] They found that performance on two different tests of creative thinking was inversely correlated with the structural integrity of white matter pathways in the left and right inferior frontal lobes; these pathways are part of a radiation of white matter from the thalamus. Similar to the decrease in dopamine receptor concentration in the thalamus, these types of changes overlap with changes seen in certain kinds of psychopathology (schizophrenia and bipolar disorder). Jung and colleagues do not suggest a causal mechanism for why these sorts of changes to the white matter are associated with increased creativity in healthy individuals, but they point out that optimal brain functional anatomy may encompass a range of physiological states, working in conjunction to maintain distinct but overlapping cognitive processes.

Although it is early days in this research, the functional neuroanatomy of the creative brain is likely to involve many regions in both hemispheres, depending on the task and the context in which that task is carried out. Certain networks will undoubtedly be more important than others in terms of creativity monitoring and motivation, but because the set of problems requiring creative solutions in the real world is more or less unbounded, it is likely that the human brain produces creativity based on its ability to draw on all of its diverse cognitive resources. Creative people will vary, by biology or training or both, in terms of which cognitive networks they use to be more creative in their chosen areas. As a species, we benefit from the diverse ways in which creativity can emerge.

A Creative Kitchen Climate

Some level of creative expression should be within the reach of everyone's brain. Whether or not a person achieves his or her creative potential depends on the environment in which he or she lives and works. Accordingly, business and management psychologists have investigated the working climate that best nurtures the creative inclinations of workers.[28] The conditions that make for a creative kitchen are undoubtedly similar to those that are successful in other creative working environments. Business psychologists have found that working climates that provide a challenge, intellectual stimulation, a sense of mission, and positive collegial relationships encourage creativity and innovation.[29] Support from management in the form of resources and autonomy are also important variables, sending the message that creative work is both nurtured and valued.[30]

Running a high-level restaurant is a creative business, or at least it should be. There are two main ways that a fancy restaurant can express its creativity: via its decor and via its food. It is always somewhat distressing, at least to food-oriented people, when a restaurant seems to care more about the decor than the food. On the other hand, no one wants to eat in a place that does not meet a certain standard of comfort, service, and hygiene. A restaurant's decor is not the primary indicator of good, creative food, but it is a signifier to both the patron and the cooking staff that the pursuit of creativity may be supported by the people who write the checks.

Researchers Véronique Chossat and Olivier Gergaud wondered what aspects of creativity were being rewarded by restaurant critics in France.[31] Critics are an essential part of the economy and culture of haute cuisine. Their opinions shape where restaurant patrons choose to spend their time and money; in turn, critics' opinions reflect to some extent the prevailing view of those consumers. Chossat and Gergaud systematically analyzed reviews of restaurants in one of the most popular restaurant guidebooks in France, *Gault-Millau*. They chose only the restaurants of chefs who were members of an elite association, the Master Chefs of France (185 in total). In effect, their goal was to determine whether Alain Ducasse, one of the most famous French chefs, would get the same high ratings in a roadside cafe as in a luxury restaurant. Chossat and Gergaud found that while a restaurant's appearance had an important effect on whether reviewers classified it as a top restaurant, this factor was not as important as the quality and creativity of the food. Restaurant entrepreneurs may expect some return on their investment in nice decor or an extensive wine list, but if they neglect the creative work in the kitchen, then they may be at risk of losing everything. So Alain Ducasse probably would get a pretty good review if he worked in a roadside cafe, although it would be a better one if the surroundings were a bit more upscale.

So it is the creativity that goes on behind the closed doors of the kitchen that matters most. In their cookbooks, Adrià, Keller, and Achatz, our trio of notable über-creative chefs, all make it clear that their cooking happens in the context of a team, a family even, of supportive and dedicated staff.[32] In his book, Achatz includes a photograph in which no fewer than sixteen assistants are seen working, each at his or her own station, in a pristine kitchen; in that photo, Achatz himself is in a corner, illuminated by sunlight streaming in from the only visible window, conversing with an associate. Keller includes a photograph of a staff meal, apparently featuring himself as the cook (the photo is a bit blurry, so it is hard to be sure), although he states that "the staff meal cook is a low man in a kitchen hierarchy." But he also believes that a cook who can be passionate about making the staff meal, pushing his or her own abilities and imagination in the limited setting, is someone who potentially has the makings of a "great chef." Adrià features dozens of photos of his staff, whether eating their staff meal or preparing dinner in "the kitchen at full speed." His kitchen staff is portrayed as more than a family—it is a full-blown creative commune.

All three of these chefs have established workspaces that clearly serve to maximize creative input and output. Just having a large number of assistants does not make a kitchen a creative environment, of course. Traditional haute cuisine restaurants had a whole brigade of staff, arranged into a rigid hierarchy. The modern creative chef takes the brigade and turns it into a creative force. Research by Jeou-Shyan Horng and colleagues on chef creativity shows that an innovative kitchen requires teamwork, especially for the discussion and exchange of ideas and techniques.[33] Although some head chefs are reluctant to share all of their techniques and ideas with underlings, this may ultimately be counterproductive. Based on interviews with chefs, Horng finds that they often have creative insights after brainstorming ideas with colleagues and staff. In this regard, one value of a large staff is that it may act not simply as the hands of the executive chef but maybe a little bit as the head as well.

Of course, Adrià, Keller, Achatz, and others at their somewhat stratospheric level provide something else that all of creative management techniques in the world cannot easily conjure up: a charismatic leader as a literal and figurative source of inspiration. Chefs of this kind are more like symphony conductors who also happen to be composers. They direct a group of creative individuals to fulfill their own particular visions. For the most part, the various assisting chefs must toil in anonymity, except perhaps for the sous chef, the concertmaster of the restaurant kitchen. Ultimately, a subordinate chef must decide when the time is right to get out from under the wing and reputation of the head chef.

The business psychology literature on creativity is implicitly focused on the nurture side of the nature/nurture equation in terms of the origins of creativity. After all, such literature is dedicated to improving working environments to enhance the potential creativity of all workers. But it is clear that no matter how nurturing of creativity a kitchen environment might be, there are chefs of special talent, or genius, who transcend the normal levels of creative vision and performance. Of the thousands of people who enter commercial kitchens with the hope of making it as a great chef, only a select few have the combination of talents necessary to succeed at the highest level. In considering the elite restaurant kitchen as a workplace, it becomes apparent that one of the talents a great chef must have is leadership. The word *chef*, from a French word with the root meaning of "head," implies this, but that is easy to overlook as a historical convention, a relic of a different time. Still, for countless millennia one of the most fundamental aspects of human behavior has been our ability to work together in small, face-to-face groups toward complex common goals. Leading such groups requires a delicate mix of skills, the ability to coerce or cajole as needed. In order to create a truly creative workplace, a great chef needs these sorts of skills, along with the sensory imagination to think of new ways of eating.

The Other Creative Kitchen

The January/February 2011 issue of *Saveur* magazine ("Savor a World of Authentic Cuisine") was dedicated to one of their standard features: an annotated and illustrated list of 100 food-related items. In this case, the recommendations and tricks of the trade were solicited from 76 chefs. As the introduction described them: "Chefs are a special breed. They're dedicated artists who live and breathe food. Some are ambassadors of international cuisines; others wow us with their creative interpretations. All are teachers who inspire us to become the best cooks we can be." They are also overwhelmingly male—well, at least among the 76 that *Saveur* surveyed: 62 were male and 14 were female, a ratio of about 4.5 to 1. An even higher ratio of men to women can be found among the more than 200 chefs identified as the year's "Best New Chefs" between 1988 and 2010.[34]

Actually, in historical terms, today there are probably an unprecedented number of women being recognized as chefs at the highest level. Of course, women have long been well represented as cooks in the food service industry, especially in large institutions such as schools and hospitals. Women have also been chefs in restaurants for generations, especially in family-owned establishments. The home economics field was dominated by women in the twentieth century, and after World War II many women were employed in various industrial test kitchens. Women such as Fannie Farmer and Irma Rombauer wrote the cookbooks that became home cooking bibles for generations.

In cultures throughout the world, the domestic side of food preparation has typically been the domain of women. Although celebrated at an emotional level, home cooking is typically thought of as traditional and conservative, and perhaps in the modern media age it is seen as more imitative than creative. Like textile arts such as quilting or knitting, the culinary creativity of domestic women has largely been unpaid and anonymous. One reason for this is that how innovations are developed and propagated privately is totally different from how it is done in the public, male-dominated world of creative commerce. Domestic creativity is anonymous in part because it is social and collaborative, enmeshed in informal but important personal relationships. Creativity is often collaborative in the public sphere, but the collaborative structure is formal and hierarchical, so credit is given to whoever is at the top (e.g., the executive chef).

The conflict between private and public kitchen creativity has been explored by Lisa Heldke in an analysis of recipe sourcing for "ethnic" cookbooks.[35] Heldke focuses on Claudia Roden's *A Book of Middle Eastern Food* (1974), a pioneering cookbook noted for Roden's efforts to personally track down and record recipes from the women who used them on an everyday basis. Heldke writes: "The unpublished women who contributed recipes become interchangeable parts, relevant only for the (universalizable) quality of their being 'native cooks.'"[36] Heldke's critique could be extended to any number of ethnic cookbooks or recipe collections that seek to assemble and record traditional knowledge. Such an activity is laudable in many ways; doing field work and interviewing dozens of women to get the information right is certainly preferable to relying on a select few published sources or key informants. The ironic thing, as Heldke points out, is that once Roden published her cookbook, the age-old recipes became "hers" from a legal and publishing perspective, and should another cookbook writer use her recipes without attribution, she might feel she was being robbed.

Heldke does not accuse Roden and others like her of stealing. But in the process of creating one cultural product from the cultural products of others, traditional knowledge becomes a commodity to be mined. In a sense, recipes become not unlike other traditional products that wind up in museums or stores, a symptom of a colonial system in which traditional knowledge is available (i.e., unowned) simply because its originators worked anonymously and collectively.

Home cooking in developed countries is similar to traditional ethnic cooking in the sense that there is no ownership of recipes and creative innovation is largely uncredited. Starting in the 1950s, home cooking in the United States underwent many changes, including the introduction of new technologies and products, which in turn reflected fundamental changes in the household. Creative innovations were introduced from above as the home economists and food scientists of many industrialized food concerns sought to bring efficiency and convenience to American kitchens. Jean Anderson writes: "Many of the recipes our mothers and grandmothers loved were product-driven: canned-soup casseroles, molded salads, mayonnaise cakes, graham cracker crusts, even chocolate chip cookies."[37] All these innovations had innovators behind them, of course, many of them women, but the low-key introduction of an anonymous recipe via a can label or a magazine article was probably a more "organic" and social way to spread domestic innovations.

One of the ways that "ladies" (as they were undoubtedly addressed) could get some recognition was if they got together as a group and assembled a cookbook of home recipes, usually to support a charity or cause. These were usually printed privately, often with a spiral binding or in a ring binder. I have one of these in my collection called *Guten Appetit from Amana Kitchens.* The Amana villages are a tight-knit group of communities in eastern Iowa that began as a German Christian socialist collective in the nineteenth century, remaining completely self-sufficient until the 1930s.[38]

Despite being the product of a somewhat more exotic than usual American cultural history, *Guten Appetit*, first published in 1985, seems a fairly typical example of a do-it-yourself collective cookbook. There is a mixed collection of soup-can casseroles (reflecting the era when most of these cooks came of age), traditional German specialties, sweets, and in keeping with their rural setting, game. I am taken with the simplicity of "Apple Squirrels," which is basically squirrels stewed in water and cider vinegar; adding Bisquick dumplings is optional. Facsimiles of signatures are attached to each recipe. This is interesting from the standpoint of claiming ownership. Many of these recipes are clearly traditional or from other sources; the truly original and creative ones are not distinguished from those that are merely conveyed. Looking through the recipes, it is a bit hard to discern which ones might be truly creative, but the range of dishes suggests a reasonably adventurous palate (and a fondness for coffee cake, for which there are six recipes).

When it comes to food preparation, it becomes quite clear that in a cultural sense, two separate creative domains have evolved: one private (or domestic), anonymous, and dominated by women, and one public, commercial, and male-dominated. When we think of creative achievement in cooking, the focus is usually on the latter of these two domains. In pre-feminist days, a man interested in cooking could always deflect criticism or ridicule by pointing out that "all the best chefs are men." It is no secret that the commercial restaurant kitchen has historically been dominated by men, who have created in them an atmosphere that has often been characterized as "testosterone-fueled." The tremendous bias toward men in recognizing creative achievement in the kitchen has, of course, historically been seen in almost all other public creative fields (acting being a notable exception).

Before the 1970s, the number of female elite chefs who would be recognized in a magazine such as *Saveur* or *Food and Wine* would have been vanishingly small. There has been an obvious trend toward more women achieving such recognition (something under 20 percent) in recent decades, compared to the mid-twentieth century. But has a ceiling been reached? Data for the *Food and Wine* best new chefs are available back to 1988. Between

1988 and 2002, typically two or three of the ten best new chefs were women ("new" here basically means someone who has run a restaurant kitchen for not more than five years). From 2003 through 2010, however, the average has been only one woman among the ten best new chefs (2003 had zero, and 2008 had two). So if there was a trend toward greater recognition of women for approaching the highest level of culinary creativity—and recognition such as this is critical in a developing career—then that trend seems to have reached a plateau or even reversed.

This gets us to a basic question: why are there not more women working as elite creative chefs? We have now had several working generations of women go through the system. Some have had great success, but the overall numbers still do not approach those of men. The work environment of the restaurant kitchen has something to do with it: more men than women are comfortable working in it. This may be changing, but social change can be slow. Also, the work hours clearly play a role. As in other professions, young women often have more difficult choices to make than young men when it comes to balancing commitments to work and family. Work hours at fancy restaurants are not amenable to great flexibility; the hours when restaurants can make money are relatively limited. These are perfectly understandable reasons for why the opportunities for women in creative restaurant work might be limited.

But maybe women do not succeed as highly creative chefs because in general there are more creative men than women, at least for the set of talents it takes to be a great executive chef. If this is the case, and if the restaurant kitchen is also generally culturally less hospitable to women, then that would be a double whammy against women being elite chefs. Is there any evidence that women are less creative than men?

The strongest source of evidence that men are on average more creative than women is the public record of creative achievement, but everyone acknowledges that historically the opportunities for women to achieve creatively have been severely limited compared to those for men. As women have been allowed to work in creative fields in greater and greater numbers, there have been increases in their representation among the ranks of those widely acknowledged for their creativity. I am not aware of any studies of brain function that suggest that men and women have significant differences in their ability to be creative. However, there is a significant, if not overwhelming, psychological literature on gender and creativity.

This literature, based on the many tests that psychologists have devised to measure creativity under controlled conditions, does not suggest that creativity is a gendered behavior. In a major review of this field, John Baer and James Kaufman concluded, "Lack of differences between girls and boys, and between men and women, is the most common outcome of the many studies. . . . [I]t should be noted that if there were to be an overall 'winner' in the numbers of studies in which one gender outperformed the other, it would be women and girls over men and boys."[39] And yet, as Baer and Kaufman point out, these tests have generally not been predictive of creative achievement in the real world, where men significantly outperform women. They conclude that at this point environmental differences are probably the best explanation for gender differences in creative achievement, especially since there do not appear to be any strong differences at a more basic cognitive level.

Baer and Kaufman point to "aptitudes, motivations, and opportunities" as factors essential to explore if we are to understand gender differences in creative achievement in the real world. Of course, part of understanding this type of achievement depends on one's definition of the "real world." Creative achievement has typically been measured or considered in the public sphere, where it can be attributed to the actions of a specific individual. Creativity, of course, happens at all different levels of society. In fact, as we've seen, private, small-scale, face-to-face, collective creativity is really what people are all about, at least in evolutionary terms. What we recognize and celebrate as great creative achievements in a large-scale, industrialized, media-saturated society may just be the tip of the iceberg. Now, it may be a glorious, awe-inspiring, magnificent iceberg tip, but it should not blind us to the creativity that goes on all around us every day in small ways. Perhaps this is creativity with a small c compared to the capital-c creativity of an Adrià or an Achatz, but we know that over the vast time scale of evolution, even small trends can have major effects given enough time.

Creative Choices

For much of human history, being creative about food was not a luxury but a necessity. Once we became superomnivores, finding new foods and new ways to eat old foods developed into an essential strategy for survival. Although scarcity might exert the ultimate selective pressure to be creative about food, it is also likely that the kinds of creativity involved in feasting, when times were good, could have been just as important. The social cohesion and harmony fostered by sharing an abundance of food in a symbolic setting pays evolutionary dividends in many ways. The feast is a form of creative expression, not just as it is seen in the great civilizations of Rome or China but as it occurs in any group of people who designate a meal as special and connected to greater themes, concepts, and ideologies.

In the modern food environment of developed countries, where many people have ready access to relatively cheap and abundant food, there is no particular reason to be creative about eating. Scarcity is not an issue, and even the domestic creativity of stretching a pricey ingredient to go a little further is rarely valued. So food creativity is a choice, in terms of either spending money for it in a restaurant or producing it in the home. Why, then, do people seek out creativity in food? Look back at the quote from Thomas Keller at the beginning of this chapter. He puts his finger on a key aspect of the creative experience—that it should be "satisfying" and promote "feelings" and "passions." It should come as no surprise that the dopamine reward/motivation system is involved in creativity: people can become "addicted" to creative acts the same way some are to chocolate. For those not so addicted, creativity can still bring pleasure, just like eating a pleasant food can. And in the same way that we are motivated to seek the sweet or fatty, we may also be motivated, by both our biology and culture, to be creative.

8

THEORY OF MIND, THEORY OF FOOD?

Animals feed themselves; men eat; but only wise men know the art of eating.

The pleasures of the table are for every man, of every land, and no matter what place in history or society; they can be a part of all his other pleasures, and they last the longest, to console him when he has outlived the rest.

—JEAN ANTHELME BRILLAT-SAVARIN, *The Physiology of Taste*, trans. M. F. K. Fisher (Alfred A. Knopf, 2009)

I HAVE SO FAR MANAGED TO AVOID drawing from the well of Brillat-Savarin's gastronomical aphorisms, but I could resist no longer. Although Brillat-Savarin may be the epitome of a very French sort of gastrophilosophy, he makes it clear that he sees the art and pleasures of eating as a human universal. The universal importance of food and eating should be obvious to everyone. What we eat, how we eat, and even why we eat—all of this comes effortlessly to us. At least that is our perception. Knowledge and habits associated with food and eating come to us as naturally as breathing. Given how important food is to basic survival, this is not surprising. We are animals, and in order to do all the things that animals need to do, we need to eat.

We have a tendency to take for granted the complex cognitive tasks that come to us naturally as we grow up. As a result, the complex neural machinery underlying such abilities remains largely hidden from us. Compare the unthinking ease with which we acquire our first language to the gear-grinding difficulty that many of us have in learning a second or third language. The underlying neural basis of first-language acquisition and ability is far from simple, but our brains are programmed by natural selection to acquire this skill without effort as we grow up, at least for the vast majority of people. In contrast, learning a second language at an age well beyond the critical period of early childhood is not so easy, and in the process many of us become fully aware of the cognitive tasks (e.g., speaking and memory, among others) involved in becoming fluent.[1]

Adopting a new diet can be similar to learning a second language—it can be hard to do. It is not impossible, of course: people successfully change their diets all the time. But more often than not they wind up going back to the way they used to eat. Changing one's diet can be difficult on many different levels. Compared to the way we eat naturally—that is, the diet we acquire as we grow up, eating certain foods in certain contexts—learning a new diet can be like learning a new language. Our brains are wired for the diets we learned at the family table, or wherever exposure to food takes place as children grow up in different cultures or times. This diet becomes our normative or cognitive diet—the diet that has no name, as we discussed in Chapter 6.

A complex cognitive task such as language involves a distributed network of brain regions. Certain regions stand out more than others as being critical for language (such as Broca's area for the motor control of speech), but all of them must work in concert to produce what a listener would describe as normal speech. A fascinating thing about language is that the environment determines the form—the specific languages—that are produced. The ability to speak and understand language has been strongly selected for over the course of evolution. Yet the language networks in the brain can work with an apparently infinite variety of languages (and expand into nonspoken realms, such as sign language and writing).

Complex cognitive tasks are best thought of as both biological and cultural, in that they may reflect biological hardwiring, but that wiring is typically laid down with a cultural blueprint. The normative diet we have in our minds—the cognitive model for food and eating—is, like language, an example of how natural selection shapes our neurobiology to respond to cues in the environment to produce an adaptive cognitive process. Now, I do not think that language affords the only or best model for the cognitive basis for the diet we carry in our minds. Spoken language is an external behavior, observable by others and tested in social interaction. Our cognitive diet is an internal model of objects in the world that we classify as food. I suggest that we all have a "theory of food" that reflects the internalized state within our minds governing our relationships and interactions with food. It develops as we grow up in a certain cultural environment, and becomes an adaptive part of our adult cognition. However, the way the theory of food shapes an individual's food thinking has in itself been shaped by evolution to occur in a certain kind of environment. The environment of the developed world is not necessarily that kind of environment.

Theory of Mind

If language is not a totally appropriate model for theory of food, then a better one may be found in the influential hypothesis known as "theory of mind" (ToM). Comparative psychologists David

Premack and Guy Woodruff introduced the ToM concept in 1978.[2] Premack and Woodruff were interested in comparing the cognition of chimpanzees and humans in terms of their ability to predict or estimate or impute the mental states of others. Theory of mind provided them with a means of comparing cognition across species.

What exactly is theory of mind? Consider yourself in your role as a social actor. How do you know what other people might want or believe, or whether or not they are behaving "for real" or pretending? Premack and Woodruff argued that in order to be able to function in an interactive social group, especially one composed of such socially complex creatures as human beings, an individual needs to have an implicit theory about the mental states of others. A more formal way of saying this, according to Alan Leslie, is that people are "endowed with a representational system that captures the cognitive properties underlying behavior."[3] Premack and Woodruff said that the "theory" part of ToM is in recognition of two facts: that it reflects a state of mind that is not observable by outsiders, and that it is used by individuals to predict the behavior of others. Theory of mind is a complex cognitive ability that has been shaped by natural selection in response to the complex, interactive social environments that have arisen over the course of human evolution.

Like language, theory of mind appears to undergo a fairly predictable process of development in children, with its sophistication increasing as age increases. Numerous tests have been developed to examine different types and levels of ToM ability in children. One of the classic scenarios to test ToM was developed by Leslie and his colleagues. It goes like this:

> Sally has a marble that she places in a basket and covers, and then departs. While she is gone, Ann removes the marble

from the basket and places it in the box. A child [the research subject] to whom this scenario is presented is asked to predict where Sally will look for the marble when she returns. To correctly predict Sally's behavior, it is necessary to take into account both Sally's desire for the marble and Sally's belief concerning the location of the marble. In this scenario, Sally's belief is rendered false by Ann's tampering. Therefore, to succeed on this task, the child must attribute to Sally a belief that, from the attributer's point of view, is false.[4]

Children as young as four years of age have no problem discerning a false belief in scenarios of this kind. Even by the age of two years, children are quite expert at determining if another individual is pretending. For example, a two-year-old child seeing her mother speak into a banana as though it is a telephone knows that the mother is pretending. This knowledge reflects the child's theory of mind (accurate in this case) about the mother's mental state.

Theory of mind has become an important focus for research in many fields. In psychiatry, it has been widely used to assess social function in conditions such as schizophrenia and especially autism. Research by Simon Baron-Cohen and colleagues indicates that theory of mind deficits emerge in autistic children at eighteen months or even earlier. Baron-Cohen likens autism to a kind of "mind-blindness."[5] People with autism or Asperger syndrome are unable to mind-read: they cannot imagine the thoughts and feelings of others. Thus, as Baron-Cohen writes, "they find other people's behavior confusing and unpredictable, even frightening."[6]

The presence of ToM deficits in autism suggests that a neurobiology of ToM exists, and that in certain disorders it is not working correctly. Unfortunately, research on autistic individuals has not led to the identification of a specific affected region or regions. In fact, Sarah Carrington and Anthony Bailey, in a recent review of ToM neuroimaging studies, point out that many regions seem to be active during the variety of ToM tasks that researchers have investigated.[7] Parts of the frontal lobe (the medial prefrontal cortex/orbitofrontal regions) and the superior temporal lobe are activated most often in these studies, but not always. Like other complex cognitive functions, ToM seems to be dependent on a widely distributed, overlapping complex of neural networks.

A model proposed by Marcel Adam Just and Sashank Varma attempts to account for the dynamic brain processes that underlie forms of complex cognition, such as ToM.[8] They start with a basic principle, which is generally agreed upon by all cognition researchers: "Thinking is the product of the concurrent activity of multiple brain areas that collaborate in a large-scale cortical network."[9] According to Just and Varma, these cortical networks change according to the demands of the thinking-related task. Although there is some specialization in the cortex, cortical areas can generally perform multiple functions, and different functions can be performed in multiple cortical areas. This flexibility allows not only for the reformation of networks following brain injury but also for the dynamic recruitment of regions to attend to tasks on a regular basis. Although the brain may develop primary neural networks for specific complex cognitive tasks, variations on a theme are possible, due to the recruitment of different cortical areas toward similar—but not identical—goals.

Cognitive researchers are just beginning to get a handle on the nature of complex cognition in the brain, both at the experimental level of cognitive neuroimaging and at the more theoretical level of modeling how the brain thinks. The ToM concept, which has its origins in the comparative, evolutionary study of behavior, demonstrates that what goes on between our ears is shaped by how we have responded to the cognitive demands of our ancestral environments. For humans and other primates, the social dimension is one of the most critical aspects of these environments.

Before moving on, let's take a quick look back at the question that got ToM research started in the first place. Premack and Woodruff wondered, "Does the chimpanzee have a theory of mind?" Their own studies in the late 1970s suggested to them that chimpanzees could observe the actions of others, discern motives or goals, and then act in a manner consistent with those perceived goals. They thought that the chimpanzees had a ToM, although not necessarily identical to the one in humans. However, some later researchers, using a variety of other ToM tasks, maintained a certain level of skepticism about this claim.[10]

In 2008, Josep Call and Michael Tomasello reviewed thirty years' worth of research on chimpanzee ToM and came to the conclusion that chimpanzees definitely do have a working ToM.[11] Chimpanzees can clearly understand that a human actor (as is often presented in experiments of this kind) has intentions or goals, and can discern whether or not an actor they are observing possesses knowledge about a situation (e.g., via seeing or hearing something). One thing that chimpanzees cannot recognize, but which human children at a fairly young age can, is when an actor possesses a false belief (the kind of task tested in the Sally/Ann scenario above). Call and Tomasello argue that the only way it can be said that chimpanzees do not have theory of mind is if it is taken to be equivalent to the ability to recognize false beliefs. This is clearly too limiting, and Call and Tomasello conclude that "chimpanzees understand how . . . psychological states work together to produce intentional action."[12] Chimpanzees understand, at least to some degree, the underlying motivations and perceptions that underlie the behavior they observe in others. For most researchers, this means they possess a theory of mind.

Theory of Food

I said at the outset that this book would be as much about *thinking* food as eating food. I suggest that one of the main ways that we think about food is with an implicit theory of food (ToF), an internal, cognitive representation of our diets in our minds. My hypothesized ToF is analogous to theory of mind, and shares with it many of the basic features of complex cognition we discussed in the previous section. Theory of mind evolved because humans (and other primates) live in highly interactive social groups that place a premium on the ability to read the minds of other social actors. Similarly, our ToF evolved not only because food is important for survival and we must learn how and what to eat as we grow up, but because our complex language-based cultural environment embeds food in an extensive web of other cognitive associations.

All primates to some extent learn how to eat as they grow up, observing what their mothers and other members of their social group do with food items. Theory of food, like theory of mind, is not necessarily unique to humans. But I believe that, as with ToM, our sociocultural environment and enhanced cognitive abilities take ToF to a level that is not seen in other animals. Our ToF cannot be just about nutrition, because when it comes to the most important and critical aspects of our survival, the boundaries between the merely physiological and the cultural become blurred. In the same way that the sexual dichotomies of biology become the continuum of gender under culture, and sexual reproduction goes from being an act between a male and a female to a social institution, food and eating are much more than ingestion and digestion. We need a ToF not simply for sustenance but also to understand one of the basic currencies of human social existence.

Similar to ToM and language, the form of an individual's ToF is likely shaped during a critical period in childhood. Developmental psychologists have long charted the normal progression of diet development from infancy through weaning and the transition to more adult "table foods." This period has been undergoing increased scrutiny with the onset of the obesity epidemic in the developed world. The importance of the food environment during this developmental stage for establishing lifelong food habits is highlighted by the fact that researchers are targeting this period for potential obesity interventions. Leann Birch and colleagues suggest that new parents, especially those who are overweight, should be taught feeding strategies for their infants that would serve to potentially lessen the risk for the child developing obesity later.[13] These include teaching the parents soothing strategies other than feeding, and recognizing other forms of distress besides hunger.

No one would doubt that the food and eating habits developed in childhood can strongly influence adult patterns. These habits are in part the external manifestations of the broader ToF that begins to take shape in childhood. However, theory of food goes beyond the observable food habits of an individual: it encompasses how food is thought and what it means.

In the previous chapters, we have discussed how many different cognitive processes relate to the human experience of food. First and foremost for many people is the role that the senses play

264 THE OMNIVOROUS MIND

in eating. The taste, the smell, and even the feel or sound of food combine to shape the immediate eating experience. The depth and meaning of any eating experience, however, is influenced by several other cognitive processes: memory, motivation, fear of the new, contempt for the familiar, and so on. The social context of eating also always plays a role: an identical food eaten in two widely different social contexts is no longer the "same." A hot dog eaten at a baseball park would not be the "same" as an identical hot dog bought from a street vendor and eaten on the street while walking to a job interview. The difference arises because our internal ToF takes into account more than just the sensory or nutritional content of a food.

Like theory of mind, an individual's theory of food is necessarily shaped by both genetic and environmental factors. The brain is tremendously plastic in terms of what it can do, but there are always constraints imposed by biology; similarly, biology confers on some individuals unique potentials that can be unleashed given the opportunity. Theory of food, like all other forms of complex cognition, will therefore vary among individuals because of both genetic and environmental factors. Some people will have a ToF that makes them more comfortable with a limited diet of familiar foods, while others will range much more widely, using food as a vehicle for exploration and adventure.

As for theory of mind, it will probably be impossible to pinpoint a single brain neural network that accounts for theory of food. Since theory of food can be as much about not eating as about eating, we would not even predict that those parts of the brain associated with ingestion should form a sort of default network. The absence of a single dedicated neural network involving regions x, y, and z in a predictable sequence does not mean that there cannot be an evolved propensity for developing a ToF.

Complex cognitive processes such as language and ToM are undoubtedly adaptations, and our ability to form a ToF is also very likely adaptive. We are in the early days of understanding the biological basis of complex cognition. Experimental methods have necessarily favored understanding them via the separate components of individual networks rather than as an intact operating system involving multiple, interacting networks.

Theory of food should be strongly influenced by the developmental environment in the same way that a child learns the language that he or she hears while growing up. But is there anything biologically deeper that might be expressed in ToF? Consider again that many of our behavioral and anatomical adaptations evolved in an environment quite different from the urbanized, developed environments of the present day. People living in developed countries today have ready access to foods rich in fat or sugar, already largely prepared, and eaten not so much according to hunger as in adherence to a set schedule. Eating is often more solitary than social, and most people are isolated from the natural sources of food. Seasonality in eating is largely nonexistent— not only are toaster pastries, diet soda, and potato chips always available, but so are tomatoes, asparagus, and citrus fruit.

Theory of food did not evolve in a modern environment. The normal developmental environment for ToF may therefore include limitations on the amount and availability of high-quality, nutrient-dense foods, marked seasonality of foods contributing to greater variety, and periods of food shortage. Almost all individuals in this traditional environment likely had a thorough understanding of the processes of food acquisition and preparation, from the hunt or harvest to food preparation and eating. Food was more likely to represent social currency than to be bought with currency. Meals would more often be taken with members of extended kin groups than centered on just the nuclear family. Food was more connected to religion and ritual activities than it is today.

I am not sure how seasonality, familial connections, or ritual might specifically influence the development of ToF. However, taken as a whole, the modern environment is in some ways relatively impoverished compared to a more traditional environment when it comes to people's relationship with their food. It is not impoverished in the sense of calories or nutrients available; anything but. However, by choice many people only eat a small number of foods, which appeal to them largely for their fat or sugar content. In most developed countries, the social and ritual contexts of food consumption have been deemphasized. And where food is abundant, eating becomes divorced from hunger. Emotional eating, eating only to push the pleasure buttons in the brain, eating out of boredom, or eating in order to put off doing something else—these are options that would have been rare in the evolutionary past. All of these factors mean that the typical ToF a person might have today in the modern, developed world differs from a more traditional ToF not just in content but also in terms of its underlying cognitive associations.

I have gotten a bit ahead of myself, but it is worth considering some of the implications of this hypothesized ToF for people living in the current food and eating environment of the developed world. Too many people in this environment are too fat and need to lose weight; at least, that is the prevailing opinion of nutritionists, public health officials, physicians, personal trainers, and all of the other professionals who think about this issue. Losing weight generally requires dieting, and as we have already discussed, dieting can be difficult. Theory of food highlights one reason for this difficulty. Our ToF is formed as we learn about food and eating as we grow up. Adopting a new diet, in effect modifying our

ToF, is to some extent like learning a second language, except more so—it is like replacing our first language with a new one. Now, I don't think that food habits are quite as cognitively ingrained as our first language, but the two are not totally dissimilar, either. Our theories of food enmesh the things we eat in a cognitive web; the wholesale replacement of the foods that form an integral part of a ToF can have widespread cognitive ramifications, hence it is resisted. Our basic behavioral plasticity and flexibility mean that we can make adjustments to our diets. But such adjustments take time and effort, especially if the components of the old diet remain readily available.

Attitudes toward exploration and novelty should be part of ToF. It would be interesting to know whether individuals who are generally more adventurous in their eating are preadapted for dieting compared to those who eat more narrowly. Although dieting is synonymous with restraining and restricting food intake, in a more global cognitive sense it is an expansion of the baseline repertoire of eating habits. Successful dieters have a range of attitudes that distinguish them from unsuccessful dieters; a more expansive ToF may contribute to some of these more success-fostering attitudes.

Dieting is important but not exactly something that contributes to the joy of eating. So what are some of the more positive implications of ToF for food and eating behavior? One thing ToF provides us is a way to escape the hegemony of taste, texture, and satiety in our diets. There is, of course, nothing wrong with taste, texture, and satiety: we all want to eat things that taste good, are enjoyable to chew, and leave us feeling full. For countless millennia, we and our ancestors assessed potential foods based on these qualities and used these assessments to determine whether or not a specific item was good to eat. Taste, texture, and satiety were critical to our ancestors' survival and thus led to the evolution of those cognitive buttons for sweet, salt, fat, and fullness. However, these up-front qualities can overwhelm all other aspects of the eating experience. In the modern environment of easily accessible and plentiful food, eating can too easily become about consuming the crispiest, saltiest, fattiest (and often cheapest) foods until full to the gills—maybe leaving just enough room for some ice cream.

There is nothing wrong with eating foods that taste good and fill you up. However, ToF makes food much more than simply about eating, and there are sources of pleasure in food that go beyond eating. For example, we are all aware of the important connection between memory and food. As we discussed in Chapter 5, feasts such as those eaten at Thanksgiving serve as mnemonic devices for a host of events and ideologies rooted in American history. They also connect to memories (usually happy ones) involving family and friends with whom we share these meals. But we could all more enthusiastically seek out and prepare foods that bring back happy, pleasurable memories at a more personal level. We are all familiar with the "this reminds me of . . ." experience of eating, when we fortuitously happen upon a food that brings back some memory. Such memory foods, or at least the ones that bring back pleasant memories, should not be consumed only by accident. I make ketchup fried rice for my family not only because it is a delicious way to use up leftovers but also because it is something my mother made for us when I was growing up, and it is something I hope my sons make for their families someday. A particular food can be a potential catalyst for developing personal relationships, in part because the symbolic is linked to the visceral. It can be a vehicle for evoking and sustaining memories across both time and space. Memory meals do not only have to be consumed on state or religious holidays.

A sense of time also probably figures into our ToF. From acquisition to preparation to consumption, food and eating can take up a significant part of the day. Time is a valuable commodity for any primate, and too much time spent on one food can cost an opportunity to obtain another, more nutritious one. For humans, our perceptions of many foods probably carry with them an implicit calculus of how much time it would take to prepare or eat. In the age of fast food, of course, the time constraints of food need not be much of a concern, but at other times, food and eating would provide some of the significant markers of the passage of time over the course of a day.[14]

In the longer term, keeping track of food seasonality would engage not only memory but also the ability to categorize. In addition, as our ancestors developed the ability to mentally time-travel, to anticipate and discuss future events, knowledge of the seasonality of plants and animals undoubtedly became part of the strategies they used to obtain food. Seasonality influences the feeding of many animals, but for humans this seasonality could be consciously anticipated by their knowledge of the movement of the sun, the stars, and the moon, among other calendrical signs and signals. Our place in the "food-time continuum" is something that would be implicitly monitored via our ToF.

One last possible implication of the ToF is that food, food preparation, and eating are potentially sources of cognitive enhancement. The ToF implies that food is at the center of multiple brain networks encompassing several cognitive domains. Certainly the efficient function of neural networks depends to some extent on their use, as repetition strengthens the connections between neurons firing simultaneously during a cognitive task. Since ToF encompasses activity in multiple brain regions, exercising its networks should bolster or help maintain cognitive performance in diverse brain regions.

I have relied on ToM and language as analogous examples for a proposed ToF, so I will look to them again for how brain function can be enhanced via complex cognition. Some of the scientists most interested in cognitive enhancement are researchers concerned with brain aging. They are busily looking for ways to maintain cognitive performance in the brain to go along with the ways we have identified of reducing the effects of aging on the body. It is widely appreciated that physical exercise helps people maintain both body and mind as they get older. But it is also increasingly becoming clear that exercising the mind specifically is greatly beneficial for maintaining cognitive health.[15] During aging, both the continued acquisition of knowledge and engagement in cognitively challenging activities help to build up a cognitive reserve against the inevitable forces of brain atrophy and declining function (made worse by conditions such as Alzheimer's disease). There is validity to the "use it or lose it" notion when it comes to brain aging, although unfortunately it is not enough to stave off the effects of pathology indefinitely.

Maintaining positive social relationships is also important for successful aging. Both physical and mental functioning are better maintained, and the onset of dementia is delayed, in individuals who have active and meaningful social lives.[16] This is not a matter of reverse causation, where aging individuals who do better cognitively are more socially active. Longitudinal studies clearly show that social activity is a buffer against cognitive decline.

Thus when elderly people engage in intellectually stimulating activities and have a meaningful social life, they do better cognitively than if they do not. This suggests, then, that exercising the brain's language and theory of mind facilities is cognitively enhancing. Most knowledge-based activities are dependent on language, as are most (but not all) aspects of social interaction. Theory of mind, of course, is essential to social interaction. Similarly, engaging ToF in a meaningful way could also enhance cognition in the elderly. What do I mean by meaningful? It does not have to be very much; an active role in food choice and meal scheduling might be beneficial and something even quite infirm people could participate in. Continuing to be involved in the acquisition and preparation of food would be even better, as those activities involve a range of cognitive abilities. We evolved to be actively engaged with our food environments. Even more important, at an individual level the theory of food shaped by that environment is an important contributor to how the mind shapes its view of the world in general.

The importance of food at the end of life seems like an appropriate place to end this book about the natural history of food, eating, and the mind. To paraphrase Brillat-Savarin, I believe that the art and pleasure of eating belong to everyone, young and old. This is no accident of history, but the outcome of millions of years of evolution reinforced by hundreds of thousands years of cultural life. People living in the developed world, surrounded by a seemingly boundless supply of food, tend to take it for granted: food has always been there, and presumably it always will be. This attitude is unfortunate, because taking food for granted goes against both our biological and cultural heritages. Anyone lucky enough to have access to a ready and varied supply of vegetables, fruits, meats, grains, seafoods, nuts, and whatever else should celebrate that good fortune. The best way—the most human way—to do so would be to plan and prepare a nice meal and share it with family and friends.

NOTES

Introduction

1. The recipe is in E. Topp and M. Howard, *The Complete Book of Small-Batch Preserving* (Buffalo: Firefly Books, 2007), 174.

2. S. Pinker, *The Language Instinct* (New York: HarperPerennial, 1994).

3. J. Vernon, *Hunger: A Modern History* (Cambridge, MA: Belknap Press of Harvard University Press, 2007).

1. Crispy

1. Food and Agriculture Organization of the United Nations, *New Light on a Hidden Treasure. International Year of the Potato* (Rome: FAO, 2009).

2. S. Tsuji, *Japanese Cooking: A Simple Art* (Tokyo: Kodansha, 2006).

3. S. K. Srivastava, N. Babu, and H. Pandey, "Traditional Insect Bioprospecting—As Human Food and Medicine," *Indian Journal of Traditional Knowledge* 8 (2009): 485–494.

4. Ibid., 486.

5. M. Harris, *Good to Eat: Riddles of Food and Culture* (Prospect Heights, IL: Waveland, 1998).

6. L. M. Berzok, *American Indian Food* (Westport, CT: Greenwood Press, 2005).

7. C. Stanford, J. S. Allen, and S. C. Antón, *Biological Anthropology: The Natural History of Humankind*, 2nd ed. (Upper Saddle River, NJ: Prentice-Hall, 2009).

8. S. Freidberg, *Fresh: A Perishable History* (Cambridge, MA: Belknap Press, 2009).

9. Ibid.

10. J. Steingarten, *The Man Who Ate Everything* (New York: Vintage Books, 1997), 177.

11. A. J. Marshall and R. W. Wrangham, "Evolutionary Consequences of Fallback Foods," *International Journal of Primatology* 28 (2007): 1219–1235.

12. J. E. Lambert, "Seasonality, Fallback Strategies, and Natural Selection: A Chimpanzee and Cercopithecoid Model for Interpreting the Evolution of the Hominin Diet," in *Evolution of the Human Diet: The Known, the Unknown, and the Unknowable,* ed. P. S. Ungar, 324–343 (New York: Oxford University Press, 2007).

13. H. McGee, *On Food and Cooking: The Science and Lore of the Kitchen* (New York: Scribner, 2004), 778.

14. S. W. Mintz, *Sweetness and Power: The Place of Sugar in Modern History* (New York: Penguin, 1985).

15. McGee, *On Food and Cooking,* 14; S. Kawamura, "Seventy Years of the Maillard Reaction," in *The Maillard Reactions in Foods and Nutrition,* ACS Symposium Series, vol. 215, ed. G. R. Waller and M. S. Feather, 3–18 (Washington, DC: American Chemical Society, 1983).

16. McGee, *On Food and Cooking,* 304.

17. R. Wrangham, *Catching Fire: How Cooking Made Us Human* (New York: Basic Books, 2009).

18. Stanford, Allen, and Antón, *Biological Anthropology,* 8; J. S. Allen, *The Lives of the Brain: Human Evolution and the Organ of Mind* (Cambridge, MA: Belknap Press of Harvard University Press, 2009).

19. L. Aiello and C. Dean, *An Introduction to Human Evolutionary Anatomy* (San Diego: Academic Press, 1990).

20. Ibid.; J. Nolte, *The Human Brain: An Introduction to Its Functional Anatomy,* 5th ed. (St. Louis: Mosby, 2002); J. S. Allen, *The Lives of the Brain: Human Evolution and the Organ of Mind.*

21. A. Damasio, *The Feeling of What Happens* (New York: Harcourt Brace, 1999).

22. J. P. Lund et al., "Brainstem Mechanisms Underlying Feeding Behavior," *Current Opinion in Neurobiology* 8 (1998): 718–724; J. P. Lund and A. Kolta, "Brainstem Circuits That Control Mastication: Do They Have Anything to Say during Speech?" *Journal of Communication Disorders* 39 (2006): 381–390.

23. M. Onozuka et al., "Mapping Brain Region Activity during Chewing: A Functional Magnetic Resonance Imaging Study," *Journal of Dental Research* 81 (2002): 743–746; T. Tamura et al., "Functional Magnetic Resonance Imaging of Human Jaw Movements," *Journal of Oral Rehabilitation* 30 (2003): 614–622.

24. T. Takada and T. Miyamoto, "A Fronto-Parietal Network for Chewing Gum: A Study on Human Subjects with Functional Magnetic Resonance Imaging," *Neuroscience Letters* 360 (2004): 137.

25. Nolte, *The Human Brain*.

26. B. Pfleiderer et al., "Visualization of Auditory Habituation by fMRI," *NeuroImage* 17 (2002): 1705–1710.

27. N. Osaka, "Walk-Related Mimic Word Activates the Extrastriate Visual Cortex in the Human Brain: An fMRI Study," *Behavioural Brain Research* 198 (2009): 186–189; N. Osaka et al., "A Word Expressing Affective Pain Activates the Anterior Cingulate Cortex in the Human Brain: An fMRI Study," *Behavioural Brain Research* 153 (2004): 123–127.

28. L. Bidel, P. Jackson, and P. Rainville, "Brain Responses to Facial Expressions of Pain: Emotional or Motor Mirroring?" *NeuroImage* 53 (2010): 355–363.

29. J. Munzert, B. Lorey, and K. Zentgraf, "Cognitive Motor Processes: The Role of Motor Imagery in the Study of Motor Representations," *Brain Research Reviews* 60 (2009): 306–326.

2. The Two-Legged, Large-Brained, Small-Faced, Superomnivorous Ape

1. C. Stanford, J. S. Allen, and S. C. Antón, *Biological Anthropology: The Natural History of Humankind*, 2nd ed. (Upper Saddle River, NJ: Prentice-Hall, 2009).

2. C. B. Stanford, *Upright: The Evolutionary Key to Becoming Human* (New York: Houghton Mifflin Harcourt, 2003).

3. P. S. Ungar, F. E. Grine, and M. F. Teaford, "Dental Microwear and Diet of the Plio-Pleistocene Hominin *Paranthropus boisei*," *PLoS One* 3 (2008): e2044; M. Sponheimer et al., "Isotopic Evidence for Dietary Variability in Early Hominin *Paranthropus robustus*," *Science* 314 (2006): 980–982; T. E. Cerling et al., "Diet of *Paranthropus boisei* in the Early Pleistocene of East Africa," *Proceedings of the National Academy of Sciences* 108 (2011): 9337–9341.

4. J. S. Allen, *The Lives of the Brain: Human Evolution and the Organ of Mind* (Cambridge, MA: Belknap Press, 2009).

5. M. S. Springer et al., "Placental Mammal Diversification and the Cretaceous-Tertiary Boundary," *Proceedings of the National Academy of Sciences* 100 (2003): 1056–1061.

6. Stanford, Allen, and Antón, *Biological Anthropology*.

7. M. Cartmill, "Rethinking Primate Origins," *Science* 184 (1974): 436–443.

8. R. W. Sussman, "Primate Origins and the Evolution of Angiosperms," *American Journal of Primatology* 23 (1991): 209–223.

9. R. F. Kay, C. Ross, and B. A. Williams, "Anthropoid Origins," *Science* 275 (1997): 797–804.

10. K. Milton, "The Critical Role Played by Animal Source Foods in Human *(Homo)* Evolution," *Journal of Nutrition* 133 (2003): 3886S–3892S; S. B. Eaton and M. J. Konner, "Paleolithic Nutrition: A Consideration of Its Nature and Current Implications," *New England Journal of Medicine* 312 (1985): 283–289; S. B. Eaton, S. B. Eaton III, and M. J. Konner, "Paleolithic Nutrition Revisited," in *Evolutionary Medicine*, ed. W. R. Trevathan, E. O. Smith, and J. J. McKenna, 313–332 (New York: Oxford University Press, 1999).

11. S. L. Washburn, "Australopithecines: The Hunters or the Hunted?" *American Anthropologist* 59 (1957): 612–614, quote from 612.

12. R. A. Dart, "The Predatory Implemental Technique of *Australopithecus*," *American Journal of Physical Anthropology* 7 (1949): 1–38.

13. C. K. Brain, *The Hunters or the Hunted?* (Chicago: University of Chicago Press, 1981).

14. J. D. Speth and E. Tchernov, "Neandertal Hunting and Meat-Processing in the Near East: Evidence from Kebara Cave (Israel)," in *Meat-Eating and Human Evolution*, ed. C. B. Stanford and H. T. Bunn, 52–72 (New York: Oxford University Press, 2001).

15. H. T. Bunn, "Hunting, Power Scavenging, and Butchering by Hadza Foragers and by Plio-Pleistocene *Homo*," in *Meat-Eating and Human Evolution*, ed. C. B. Stanford and H. T. Bunn, 199–218 (New York: Oxford University Press, 2001).

16. H. T. Bunn and C. B. Stanford, "Conclusions: Research Trajectories on Hominid Meat-Eating," in *Meat-Eating and Human Evolution*, ed. C. B. Stanford and H. T. Bunn, 350–359 (New York: Oxford University Press, 2001),quote from 356.

17. S. B. Laughlin, "Energy as a Constraint on the Coding and Processing of Sensory Information," *Current Opinion in Neurobiology* 11 (2001): 475–480.

18. Allen, *Lives of the Brain.*

19. J. W. Mink, R. J. Blumenschine, and D. B. Adams, "Ratio of Central Nervous System to Body Metabolism in Vertebrates: Its Constancy and Functional Basis," *American Journal of Physiology* 241 (1981): R203–R212.

20. Milton, "Critical Role Played by Animal Source Foods."

21. Ibid.

22. L. Aiello and P. Wheeler, "The Expensive-Tissue Hypothesis: The Brain and the Digestive System in Human and Primate Evolution," *Current Anthropology* 36 (1995): 199–221.

23. An alternative anatomical trade-off was proposed by Karin Isler and Carel van Schaik, who looked to see if there was a trade-off between brain size and gut size in birds. They did not find any relationship between the two variables. However, they did find a trade-off between brain size and some of the muscles used in flight: birds that engage in short flights or have high flapping rates have smaller brains than those that soar or glide more. Even though muscle is not an expensive tissue metabolically, if there is enough of it, it can potentially be an important target for an energy trade-off. Isler and van Schaik hypothesized that there could have been a trade-off in hominid evolution if bipedality resulted in lower locomotor costs, which could have allowed more energy to be available to support a larger brain. K. Isler and C. van Schaik, "Costs of Encephalization: The Energy Trade-off Hypothesis Tested on Birds," *Journal of Human Evolution* 51 (2006): 228–243. See also Allen, *Lives of the Brain,* 185–189.

24. C. M. Hladik, D. J. Chivers, and P. Pasquet, "On Diet and Gut Size in Non-Human Primates and Humans: Is There a Relationship to Brain Size?" *Current Anthropology* 40 (1999): 695–697; J. L. Fish and C. A. Lockwood, "Dietary Constraints on Encephalization in Primates," *American Journal of Physical Anthropology* 120 (2003): 171–181.

25. F. H. Previc, "Dopamine and the Origins of Human Intelligence," *Brain and Cognition* 41 (1999): 299–350.

26. S. C. Cunnane and M. A. Crawford, "Survival of the Fattest: Fat Babies Were Keys to Evolution of the Large Human Brain," *Comparative Biochemistry and Physiology Part A* 136 (2003): 17–26.

27. Ibid.; M. A. Crawford et al., "Evidence for the Unique Function of Docosahexaenoic Acid during the Evolution of the Modern Human Brain," *Lipids* 34 (1999): S39–S47.

28. J. H. Langdon, "Has an Aquatic Diet Been Necessary for Hominin Brain Evolution and Functional Development?" *British Journal of Nutrition* 96 (2006): 7–17; S. L. Robson, "Breast Milk, Diet, and Large Human Brains," *Current Anthropology* 45 (2004): 419–425.

29. C. B. Stringer et al., "Neanderthal Exploitation of Marine Mammals in Gibraltar," *Proceedings of the National Academy of Sciences* 105 (2008): 14319–14324.

30. Ibid., 14320.

31. J. C. Joordens et al., "Relevance of Aquatic Environments for Hominins: A Case Study from Trinil (Java, Indonesia)," *Journal of Human Evolution* 57 (2009): 656–671.

32. P. S. Ungar, F. E. Grine, and M. F. Teaford, "Diet in Early *Homo:* A Review of the Evidence and a New Model of Adaptive Versatility," *Annual Review of Anthropology* 35 (2006): 209–228.

33. W. C. McGrew, *The Cultured Chimpanzee: Reflections on Cultural Primatology* (New York: Cambridge University Press, 2004).

34. J. Holtzman, *Uncertain Tastes: Memory, Ambivalence, and the Politics of Eating in Samburu, Northern Kenya* (Berkeley: University of California Press, 2009), quote from 94.

35. Ibid., 95.

36. P. Farb and G. Armelagos, *Consuming Passions: The Anthropology of Eating* (Boston: Houghton Mifflin, 1980).

37. Ibid., 207. Of course, extended periods of food stress or shortage can have a cumulative effect that could indeed pose a threat to the long-term survival of a group or culture.

38. A. L. Kroeber, "The Superorganic," *American Anthropologist* 19 (1917): 163–213; A. L. Kroeber, *Anthropology: Race, Language, Culture, Psychology, Prehistory* (New York: Harcourt, Brace and Company, 1948).

39. M. Verdon, "'The Superorganic,' or Kroeber's Hidden Agenda," *Philosophy of the Social Sciences* 40 (2010): 375–398.

40. P. Bellwood, "The Dispersals of Established Food-Producing Populations," *Current Anthropology* 50 (2009): 621–626.

41. G. W. Stocking, *Race, Culture, and Evolution: Essays in the History of Anthropology* (Chicago: University of Chicago Press, 1982).

42. M. Sahlins, *Stone-Age Economics* (Hawthorne, NY: Aldine de Gruyter, 1972).

43. A. S. Wiley and J. S. Allen, *Medical Anthropology: A Biocultural Approach* (New York: Oxford University Press, 2009).

44. D. Cook, "Subsistence Base and Health in the Lower Illinois Valley: Evidence from the Human Skeleton," *Medical Anthropology* 4 (1979): 109–124.

45. J. V. Neel, "Diabetes Mellitus: A Thrifty Genotype Rendered Detrimental by 'Progress'?" *American Journal of Human Genetics* 14 (1962): 353–362; J. V. Neel, "The Thrifty Genotype Revisited," in *The Genetics of Diabetes Mellitus*, ed. J. Kobberling and R. Tattersall, 283–293 (London: Academic Press, 1982).

46. J. S. Allen and S. M. Cheer, "The Non-Thrifty Genotype," *Current Anthropology* 37 (1996): 831–842; see also Wiley and Allen, *Medical Anthropology*, 96–100.

47. T. B. Gage and S. DeWitte, "What Do We Know about the Agricultural Demographic Transition?" *Current Anthropology* 50 (2009): 649–655.

48. S. B. Eaton and M. J. Konner, "Paleolithic Nutrition: A Consideration of Its Nature and Current Implications," *New England Journal of Medicine* 312 (1985): 283–289.

49. Eaton, Eaton, and Konner, "Paleolithic Nutrition Revisited."

50. S. Lindeberg, "Modern Human Physiology with Respect to Evolutionary Adaptations That Relate to Diet in the Past," in *The Evolution of Hominin Diets: Integrating Approaches to the Study of Palaeolithic Subsistence*, ed. J.-J. Hublin and M. P. Richards, 43–57 (New York: Springer, 2009), quote from 52.

51. S. Lindeberg et al., "A Paleolithic Diet Improves Glucose Tolerance More than a Mediterranean-Like Diet in Individuals with Ischaemic Heart Disease," *Diabetologia* 50 (2007): 1795–1807.

52. G. Cochran and H. Harpending, *The 10,000 Year Explosion* (New York: Basic Books, 2009).

3. Food and the Sensuous Brain

1. D. Kamp, *The United States of Arugula* (New York: Broadway Books, 2006).

2. S. Frings, "Primary Processes in Sensory Cells: Current Advances," *Journal of Comparative Physiology A* 195 (2009): 1–19; U. B. Kaupp, "Olfactory Signalling in Vertebrates and Insects: Differences and Commonalities," *Nature Reviews Neuroscience* 11 (2010): 188–200; J. R. Sanes and S. L. Zipursky, "Design Principles of Insect and Vertebrate Visual Systems," *Neuron* 66 (2010): 15–36.

3. Kamp, *United States of Arugula*.

4. M. Pollan, *In Defense of Food: An Eater's Manifesto* (New York: Penguin, 2008).

5. R. L. Spang, *The Invention of the Restaurant: Paris and Modern Gastronomic Culture* (Cambridge, MA: Harvard University Press, 2000).

6. Ibid., 146–169.

7. Ibid., 150–160.

8. Ibid., 158.

9. M. Montanari, *Food Is Culture*, trans. A. Sonnenfeld (New York: Columbia University Press, 2006), quote from 61.

10. J. A. Brillat-Savarin, *The Physiology of Taste, or Meditations on Transcendental Gastronomy*, trans. M. F. K. Fisher (New York: Alfred Knopf, 2009 [1825]), quote from 168.

11. This discussion of taste physiology is derived from D. U. Silverthorn, *Human Physiology: An Integrated Approach*, 2nd ed. (Upper Saddle River, NJ: Prentice Hall, 2001), and J. B. West, ed., *Physiological Basis of Medical Practice*, 12th ed. (Baltimore: Williams and Wilkins, 1990).

12. R. D. Mattes, "Is There a Fatty Acid Taste?" *Annual Review of Nutrition* 29 (2009): 305–327.

13. M. L. Kringelbach and A. Stein, "Cortical Mechanisms of Human Eating," in *Frontiers in Eating and Weight Regulation*, Forum of Nutrition, vol. 63, ed. W. Langhans and N. Geary, 164–175 (Basel: Karger, 2010); E. T. Rolls, "Smell, Taste, Texture, and Temperature Multimodal Representations in the Brain, and Their Relevance to the Control of Appetite," *Nutrition Reviews* 62 (2004): S193–S205.

14. G. Scalera, "Effects of Conditioned Food Aversions on Nutritional Behavior in Humans," *Nutritional Neuroscience* 5 (2002): 159–188.

15. J. Nolte, *The Human Brain: An Introduction to Its Functional Anatomy*, 5th ed. (St. Louis: Mosby, 2002).

16. Rolls, "Smell, Taste, Texture, and Temperature."

17. Ibid., S193.

18. U. Sautter, "Dining in the Dark," *Time*, July 22, 2002; R. Long, "Dining in the Dark," *AmericanWay*, March 15, 2010.

19. D. Salisbury, Dark Dining Project website, 2010, www.darkdining-projects.com/dark-dining.htm#whydark.

20. E. T. Rolls, Z. J. Sienkiewicz, and S. Yaxley, "Hunger Modulates the Responses to Gustatory Stimuli of Single Neurons in the Caudolateral Orbitofrontal Cortex of the Macaque Monkey," *European Journal of Neuroscience* 1 (1989): 53–60.

21. M. L. Kringelbach and A. Stein, "Cortical Mechanisms of Human Eating"; Rolls, "Smell, Taste, Texture, and Temperature."

22. A. Escoffier, *Memories of My Life*, trans. L. Escoffier (New York: Van Nostrand Reinhold, 1997).

23. I. E. T. de Araujo et al., "Representation of Umami Taste in the Human Brain," *Journal of Neurophysiology* 90 (2003): 313–319.

24. Silverthorn, *Human Physiology:* West, *Physiological Basis of Medical Practice*.

25. R. C. Coghill, C. N. Sang, J. M. Maisog, and M. J. Iadarola, "Pain Intensity Processing within the Human Brain: A Bilateral, Distributed Mechanism," *Journal of Neurophysiology* 82 (1999): 1934–1943.

26. C. Rennefeld et al., "Habituation to Pain: Further Support for a Central Component," *Pain* 148 (2010): 503–508.

27. D. F. Zatzick and J. E. Dimsdale, "Cultural Variations in Response to Painful Stimuli," *Psychosomatic Medicine* 52 (1990): 544–557.

28. L. Perry et al., "Starch Fossils and the Domestication and Dispersal of Chili Peppers (*Capsicum* spp. L.) in the Americas," *Science* 315 (2007): 986–988; I. Paran and E. van der Knapp, "Genetic and Molecular Regulation of Fruit and Plant Domestication Traits in Tomato and Pepper," *Journal of Experimental Biology* 58 (2007): 3841–3852.

29. P. Rozin, "Psychobiological Perspectives on Food Preferences and Avoidances," in *Food and Evolution: Toward a Theory of Human Food Habits*, ed. M. Harris and E. B. Ross, 181–205 (Philadelphia: Temple University Press, 1987); J. Gorman, "A Perk of Our Evolution: Pleasure in Pain of Chilies," *New York Times*, September 20, 2010.

30. S. Molnar, *Human Variation: Races, Types, and Ethnic Groups* (Upper Saddle River, NJ: Prentice Hall, 2006); R. J. Williams, *Biochemical Individ-*

uality: The Basis for the Genetotrophic Concept (Austin: University of Texas Press, 1979 [1956]).

31. S. Wooding, "Phenylthiocarbamide: A 75-Year Adventure in Genetics and Natural Selection," *Genetics* 172 (2006): 2015–2023, quote from 2015.

32. S.-W. Guo and D. R. Reed, "The Genetics of Phenylthiocarbamide," *Annals of Human Biology* 28 (2001): 111–142.

33. Ibid.; D. Drayna, "Human Taste Genetics," *Annual Review of Genomics and Human Genetics* 6 (2005): 217–235; B. J. Tepper, "Nutritional Implications of Genetic Taste Variation: The Role of PROP Sensitivity and Other Taste Phenotypes," *Annual Review of Nutrition* 28 (2008):367–388.

34. Drayna, "Human Taste Genetics."

35. Tepper, "Nutritional Implications of Genetic Taste Variation"; B. J. Tepper et al., "Genetic Variation in Taste Sensitivity to 6-n-propylthiouracil and Its Relationship to Taste Perception and Food Selection," *Annals of the New York Academy of Sciences* 1170 (2009): 126–139.

36. N. Soranzo et al., "Positive Selection on a High-Sensitivity Allele of the Human Bitter-Taste Receptor *TAS2R16*," *Current Biology* 15 (2005): 1257–1265.

37. S. Wooding et al., "Natural Selection and Molecular Evolution in *PTC*, a Bitter-Taste Receptor Gene," *American Journal of Human Genetics* 74 (2004): 637–646.

38. Ibid.

39. J.C. Wang et al., "Functional Variants in *TAS2R38* and *TAS2R16* Influence Alcohol Consumption in High-Risk Families of African-American Origin," *Alcoholism: Clinical and Experimental Research* 31 (2007): 209–215.

40. V. B. Duffy, "Variation in Oral Sensation: Implications for Diet and Health," *Current Opinion in Gastroenterology* 23 (2007): 171–177, quote from 173.

41. Y. Hasin-Brumshtein, D. Lancet, and T. Olender, "Human Olfaction: From Genomic Variation to Phenotypic Diversity," *Trends in Genetics* 25 (2009): 178–184.

42. H. Kaplan et al., "A Theory of Human Life History Evolution: Diet, Intelligence, and Longevity," *Evolutionary Anthropology* 9 (2000): 156–185; C. Panter-Brick, "Sexual Division of Labor: Energetic and Evolutionary Scenarios," *American Journal of Human Biology* 14 (2002): 627–640.

43. C. B. Stanford, *The Hunting Apes: Meat Eating and the Origins of Human Behavior* (Princeton: Princeton University Press, 1999). Quote from p. 200.

44. M. F. K. Fisher, *The Art of Eating*, 50th Anniversary Edition (Hoboken, NJ: Wiley, 2004). Quote from p. 584.

45. C. Lévi-Strauss, *The Raw and the Cooked* (Chicago: University of Chicago Press, 1983 [1969]), quote from 269.

46. K. Shopsin and C. Carreño, *Eat Me: The Food and Philosophy of Kenny Shopsin* (New York: Alfred A. Knopf, 2008), 91.

47. www.urbandictionary.com.

48. Y.-C. Chuang et al., "Tooth-Brushing Epilepsy with Ictal Orgasms," *Seizure* 13 (2004): 179–182.

49. J. R. Georgiadis et al., "Regional Cerebral Blood Flow Changes Associated with Clitorally Induced Orgasm in Healthy Women," *European Journal of Neuroscience* 24 (2006): 3305–3316; J. R. Georgiadis et al., "Brain Activation during Human Male Ejaculation Revisited," *NeuroReport* 18 (2007): 553–557; J. R. Georgiadis et al., "Men versus Women on Sexual Brain Function: Prominent Differences during Tactile Genital Stimulation, but Not during Orgasm," *Human Brain Mapping* 30 (2009): 3089–3101.

50. Rolls, Sienkiewicz, and Yaxley, "Hunger Modulates the Responses."

4. Eating More, Eating Less

1. B. Caballero, "The Global Epidemic of Obesity: An Overview," *Epidemiologic Reviews* 29 (2007): 1 5.

2. W. Allen, "Notes from the Overfed (1968)," in *Secret Ingredients: The New Yorker Book of Food and Drink*, ed. D. Remnick (New York: Random House, 2007). Quote from page 402.

3. E. J. McAllister et al., "Ten Putative Contributors to the Obesity Epidemic," *Critical Reviews in Food Science and Nutrition* 49 (2009): 868–913.

4. G. Taubes, *Good Calories, Bad Calories* (New York: Anchor Books, 2007).

5. D. C. Willcox et al., "Caloric Restriction and Human Longevity: What Can We Learn from the Okinawans?" *Biogerontology* 7 (2006): 173–177.

6. R. Wrangham, *Catching Fire: How Cooking Made Us Human* (New York: Basic Books, 2009).

7. M. Jones, *Feast: Why Humans Share Food* (New York: Oxford University Press, 2007).

8. W. R. Leonard, J. J. Snodgrass, and M. L. Robertson, "Evolutionary Perspectives on Fat Ingestion and Metabolism in Humans," in *Fat Detection:*

Taste, Texture, and Post Ingestive Effects, ed. J. P. Montmayeur and J. le Coutre (Boca Raton, FL: CRC Press, 2010).

9. R. D.Mattes, "Fat Taste in Humans: Is It Primary?" in *Fat Detection: Taste, Texture, and Post Ingestive Effects,* ed. J. P. Montmayeur and J. le Coutre (Boca Raton, FL: CRC Press, 2010). Mattes points out that calling fat or any other taste "primary" is a matter of definition, although sweet, sour, bitter, salty, and umami are recognized as primary tastes based on their unique and dedicated transduction mechanisms.

10. A. K. Outram, "Hunter-Gatherers and the First Farmers," in *Food: The History of Taste,* ed. P. Freedman, 35–61 (Berkeley: University of California Press, 2007), quote from 46.

11. J. E. Steiner et al., "Comparative Expression of Hedonic Impact: Affective Reactions to Taste by Human Infants and Other Primates," *Neuroscience and Biobehavioral Reviews* 25 (2001): 53–74.

12. Available at www.ers.usda.gov/Briefing/Sugar/Data.htm#yearbook (Table 50).

13. S. B. Eaton, S. B. Eaton III, and M. J. Konner, "Paleolithic Nutrition Revisited," in *Evolutionary Medicine,* ed. W. R. Trevathan, E. O. Smith, and J. J. McKenna, 313–332 (New York: Oxford University Press, 1999).

14. F. W. Marlowe and J. C. Berbesque, "Tubers as Fallback Foods and Their Impact on Hadza Hunter-Gatherers," *American Journal of Physical Anthropology* 140 (2009): 751–758.

15. G. K. Beauchamp et al., "Infant Salt Taste: Developmental, Methodological, and Contextual Factors," *Developmental Psychobiology* 27 (1994): 353–365.

16. M. L. Power and J. Schulkin, *The Evolution of Obesity* (Baltimore: Johns Hopkins University Press, 2009), 121.

17. L. Tanner, "Zoo Animals in U.S. Eating Healthier Diets," 2008. Available at www.redorbit.com/news/science/1320274/zoo_animals_in_us _eating_healthier_diets/.

18. N. Mrosovsky and D. F. Sherry, "Animal Anorexias," *Science* 207 (1980): 837–842.

19. J. J. Brumberg, *Fasting Girls: The History of Anorexia Nervosa* (New York: Plume, 1980).

20. D. A. Kessler, *The End of Overeating: Taking Control of the Insatiable American Appetite* (New York: Rodale, 2009).

21. J. Nolte, *The Human Brain: An Introduction to Its Functional Anatomy*, 5th ed. (St. Louis: Mosby, 2002); D. U. Silverthorn, *Human Physiology: An Integrated Approach*, 2nd ed. (Upper Saddle River, NJ: Prentice Hall, 2001); H.-R. Berthoud and C. Morrison, "The Brain, Appetite, and Obesity," *Annual Review of Psychology* 59 (2008): 55–92.

22. Berthoud and Morrison, "The Brain, Appetite, and Obesity."

23. E. R. Shell, *The Hungry Gene: The Science of Fat and the Future of Thin* (New York: Atlantic Monthly Press, 2002); R. S. Ahima, "Revisiting Leptin's Role in Obesity and Weight Loss," *Journal of Clinical Investigation* 118 (2008): 2380–2383.

24. S. B. Heymsfield et al., "Recombinant Leptin for Weight Loss in Obese and Lean Adults," *Journal of the American Medical Association* 282 (1999): 1568–1575.

25. M. L. Power and J. Schulkin, *The Evolution of Obesity* (Baltimore. Johns Hopkins University Press, 2009).

26. A. Wiley, *Re-Imagining Milk* (New York: Routledge, 2011).

27. T. Kelesidis et al., "Narrative Review: The Role of Leptin in Human Physiology: Emerging Clinical Applications," *Annals of Internal Medicine* 152 (2010): 93–100.

28. M. Rosenbaum et al., "Low-Dose Leptin Reverses Skeletal Muscle, Autonomic, and Neuroendocrine Adaptations to Maintenance of Reduced Weight," *Journal of Clinical Investigation* 115 (2005): 3579–3586.

29. M. Rosenbaum et al., "Leptin Reverses Weight Loss-Induced Changes in Regional Neural Activity Responses to Visual Food Stimuli," *Journal of Clinical Investigation* 118 (2008): 2583–2591. The analysis of the data from this fMRI study was a bit complicated in that there were scans done before and after weight loss and comparing subjects who had received leptin after weight loss with those who had received a placebo. For both the before and after conditions, multiple brain areas are activated upon viewing the food items, reflecting the fact that food can be a complex stimulus simultaneously activating several cognitive networks.

30. Ibid., 2587.

31. A. J. Ho et al., "Obesity Is Linked with Lower Brain Volume in 700 AD and MCI Patients," *Neurobiology of Aging* 31 (2010): 1326–1339.

32. J. S. Allen, J. Bruss, and H. Damasio, "Normal Neuroanatomical Variation Due to Age: The Major Lobes and a Parcellation of the Temporal Region," *Neurobiology of Aging* 26 (2005): 1245–1260.

33. J. S. Allen, J. Bruss, and H. Damasio, "The Aging Brain: The Cognitive Reserve Hypothesis and Hominid Evolution," *American Journal of Human Biology* 17 (2005): 673–689.

34. S. Debette et al., "Visceral Fat Is Associated with Lower Brain Volume in Healthy Middle-Aged Adults," *Annals of Neurology* 68 (2010): 136–144; S. Gazdinski et al., "Body Mass Index and Magnetic Resonance Markers of Brain Integrity in Adults," *Annals of Neurology* 63 (2008): 652–657.

35. D. Gustafson et al., "A 24-Year Follow-Up of Body Mass Index and Cerebral Atrophy," *Neurology* 63 (2004): 1876–1881.

36. Y. Taki et al., "Relationship between Body Mass Index and Gray Matter Volume in 1,428 Healthy Individuals," *Obesity* 16 (2008): 119–124.

37. N. Pannacciulli et al., "Brain Abnormalities in Human Obesity: A Voxel-Based Morphometric Study," *NeuroImage* 31 (2006): 1419–1425.

38. C. A. Raji et al., "Brain Structure and Obesity," *Human Brain Mapping* 31 (2010): 353–364.

39. A. J. Ho, "A Commonly Carried Allele of the Obesity-Related *FTO* Gene Is Associated with Reduced Brain Volume in the Healthy Elderly," *Proceedings of the National Academy of Sciences* 107 (2010): 8404–8409.

40. A. J. Ho et al., "Obesity Is Linked with Lower Brain Volume."

41. Ibid.

42. R. D. Terry and R. Katzman, "Life Span and the Synapses: Will There Be a Primary Senile Dementia?" *Neurobiology of Aging* 22 (2001): 347–348.

43. American Psychiatric Association, *Diagnostic and Statistical Manual of Mental Disorders*, 4th ed. (Washington, DC: American Psychiatric Association, 1994).

44. G.-J. Wang et al., "Evidence of Gender Differences in the Ability to Inhibit Brain Activation Elicited by Food Stimulation," *Proceedings of the National Academy of Sciences* 106 (2009): 1249–1254.

45. P. K. Keel et al., "A 20-Year Longitudinal Study of Body Weight, Dieting, and Eating Disorder Symptoms," *Journal of Abnormal Psychology* 116 (2007): 422–432.

46. L. Passamonti et al., "Personality Predicts the Brain's Response to Viewing Appetizing Foods: The Neural Basis of a Risk Factor for Overeating," *Journal of Neuroscience* 29 (2009): 43–51.

47. J. M. McCaffrey et al., "Differential Functional Magnetic Resonance Imaging Response to Food Pictures in Successful Weight-Loss

Maintainers Relative to Normal-Weight and Obese Controls," *American Journal of Clinical Nutrition* 90 (2009): 928–934.

48. E. Abrahams and M. Silver, "The Case for Personalized Medicine," *Journal of Diabetes Science and Technology* 3 (2009): 680–684.

49. T. B. Gustafson and D. B. Sarwer, "Childhood Sexual Abuse and Obesity," *Obesity Reviews* 5 (2004): 129–135.

50. Kessler, *The End of Overeating.*

51. M. J. Morris, E. S. Na, and A. K. Johnson, "Salt Craving: The Psychobiology of Pathogenic Sodium Intake," *Physiology and Behavior* 94 (2008): 709–721.

52. M. Lutter and E. J. Nestler, "Homeostatic and Hedonic Signals Interact in the Regulation of Food Intake," *Journal of Nutrition* 139 (2009): 629–632, quote from 629. See also J. A. Corsica and M. L. Pelchat, "Food Addiction: True or False?" *Current Opinion in Gastroenterology* 26 (2010): 165–169; M. L. Pelchat, "Food Addiction in Humans," *Journal of Nutrition* 139 (2009): 620–622.

53. P. Rozin, "Psychobiological Perspectives on Food Preferences and Avoidances," in *Food and Evolution: Toward a Theory of Human Food Habits*, ed. M. Harris and E. B. Ross, 181–205 (Philadelphia: Temple University Press, 1987); M. Lafourcade et al., "Nutritional Omega-3 Deficiency Abolishes Endocannabinoid-Mediated Neuronal Functions," *Nature Neuroscience* 14 (2011): 345–350.

54. M. J. Morris, E. S. Na, and A. K. Johnson, "Salt Craving."

55. P. M. Johnson and P. J. Kenny, "Dopamine D2 Receptors in Addiction-Like Reward Dysfunction and Compulsive Eating in Obese Rats," *Nature Neuroscience* 13 (2010): 635–641.

56. G. J. Wang et al., "Brain Dopamine and Obesity," *Lancet* 357 (2001): 354–357.

57. J. A. Mennella et al., "Sweet Preferences and Analgesia during Childhood: Effects of Family History of Alcoholism and Depression," *Addiction* 105 (2010): 666–677.

58. E. Stice et al., "Relation of Reward from Food Intake and Anticipated Food Intake to Obesity: A Functional Magnetic Resonance Imaging Study," *Journal of Abnormal Psychology* 117 (2008): 924–935.

59. E. Stice et al., "Reward Circuitry Responsivity to Food Predicts Future Increases in Body Mass: Moderating Effects of DRD2 and DRD4," *NeuroImage* 50 (2010): 1618–1625.

60. Some substances that we think of as drugs may be consumed in a food-like manner, thus conflating hedonic drug and food mechanisms. See R. J. Sullivan and E. H. Hagen, "Psychotropic Substance-Seeking: Evolutionary Pathology or Adaptation?" *Addiction* 97 (2002): 389–400.

61. M. R. Lowe and M. L. Butryn, "Hedonic Hunger: A New Dimension of Appetite?" *Physiology and Behavior* 91 (2007): 432–439.

62. Ibid., 438.

63. American Psychiatric Association, *Diagnostic and Statistical Manual*.

64. J. J. Brumberg, *Fasting Girls*.

65. J. E. Mitchell and S. Crow, "Medical Complications of Anorexia Nervosa and Bulimia Nervosa," *Current Opinion in Psychiatry* 19 (2006): 438–443; S. Nielsen, "Epidemiology and Mortality of Eating Disorders," *Psychiatric Clinics of North America* 24 (2001): 201–214.

66. E. Lambe et al., "Cerebral Gray Matter Volume Deficits after Weight Recovery from Anorexia Nervosa," *Archives of General Psychiatry* 54 (1997): 537–542; G. K. Frank, U. F. Bailer, S. Henry, A. Wagner, and W. H. Kaye, "Neuroimaging Studies in Eating Disorders," *CNS Spectrums* 9 (2004): 539–548.

67. American Psychiatric Association, *Diagnostic and Statistical Manual*.

68. C. M. Bulik et al., "Twin Studies of Eating Disorders: A Review," *International Journal of Eating Disorders* 27 (2000): 1–20.

69. S. Bordo, "Anorexia Nervosa: Psychopathology as the Crystallization of Culture," in *Food and Culture: A Reader*, ed. C. Counihan and P. van Esterik, 2nd ed., 162–186 (New York: Routledge, 2008 [1996]), 170.

70. W. H. Kaye, J. L. Fudge, and M. Paulus, "New Insights into Symptoms and Neurocircuit Function of Anorexia Nervosa," *Nature Reviews Neuroscience* 10 (2009): 573–584.

71. A. J. W. Scheurink et al., "Neurobiology of Hyperactivity and Reward: Agreeable Restlessness in Anorexia Nervosa," *Physiology and Behavior* 100 (2010): 490–495.

72. W. H. Kaye, J. L. Fudge, and M. Paulus, "New Insights." Quote from page 581.

73. M. N. Miller and A. J. Pumareiga, "Culture and Eating Disorders: A Historical and Cross-Cultural Review," *Psychiatry* 64 (2001): 93–110.

74. A. E. Becker, "Television, Disordered Eating, and Young Women in Fiji: Negotiating Body Image and Identity during Rapid Social Change," *Culture, Medicine, and Psychiatry* 28 (2004): 533–559; A. E. Becker et al.,

"Facets of Acculturation and Their Diverse Relations to Body Shape Concern in Fiji," *International Journal of Eating Disorders* 40 (2007): 42–50.

75. M. A. Katzman and S. Lee, "Beyond Body Image: The Integration of Feminist and Transcultural Theories in the Understanding of Self Starvation," *International Journal of Eating Disorders* 22 (1997): 385–394.

76. K. M. Pike and A. Borovoy, "The Rise of Eating Disorders in Japan: Issues of Culture and Limitations of the Model of 'Westernization,'" *Culture, Medicine, and Psychiatry* 28 (2004): 493–531.

5. Memories of Food and Eating

1. A. Damasio, *Self Comes to Mind* (New York: Pantheon, 2010).

2. L. R. Squire, "Memory and the Hippocampus: A Synthesis from Findings with Rats, Monkeys, and Humans," *Psychological Review* 99 (1992): 195–231.

3. J. R. Manns and H. Eichenbaum, "Evolution of Declarative Memory," *Hippocampus* 16 (2006): 795–808, quote from 795.

4. J. Nolte, *The Human Brain: An Introduction to Its Functional Anatomy*, 5th ed. (St. Louis: Mosby, 2002).

5. See J. S. Allen, *The Lives of the Brain: Human Evolution and the Organ of Mind* (Cambridge, MA: Belknap Press, 2009), 92–99.

6. R. Carter, *Mapping the Mind* (Berkeley: University of California Press, 1999), B. Carey, "H.M., an Unforgettable Amnesiac, Dies at 82," *New York Times*, December 5, 2008; S. Corkin, "What's New with the Amnesic Patient H.M.?" *Nature Reviews Neuroscience* 3 (2002): 153–160.

7. In a study I did with my colleagues Dan Tranel, Joel Bruss, and Hanna Damasio, we measured the size of the hippocampus in a group of patients who had experienced oxygen deprivation for various lengths of time. These anoxic events can result from carbon dioxide poisoning, a severe asthma attack, cardiac arrest, near drowning, and so on. The hippocampus is particularly vulnerable to oxygen deprivation, and anoxic patients often suffer long-term amnesia. They retain their past memories, but their ability to form new memories is severely compromised. However, some anoxic patients have few or only mild memory problems. In measuring the size of the hippocampus of these patients, we found that there was a strong correlation between the size of the hippocampus and whether or not, and to what extent, a patient suffered from amnesia. Individuals with more severe

amnesia had had more of their hippocampus destroyed during the anoxic event, while those who were better at forming new memories tended to have a more intact hippocampus. A bigger hippocampus (in the sense of retaining more of the pre-anoxia hippocampus volume) was better in a functional sense. J. S. Allen et al., "Correlations between Regional Brain Volumes and Memory Performance in Anoxia," *Journal of Clinical and Experimental Neuropsychology* 28 (2006): 457–476.

8. S. Cavaco et al., "The Scope of Preserved Procedural Memory in Amnesia," *Brain* 127 (2004): 1853–1867.

9. J. M. Fuster, "Cortex and Memory: Emergence of a New Paradigm," *Journal of Cognitive Neuroscience* 21 (2009): 2047–2072.

10. Ebert, *Life Itself* (New York: Hachette Book Group, 2011), 377–383.

11. A. Damasio, *The Feeling of What Happens* (New York: Harcourt Brace, 1999), 221.

12. K. M. Johnson, R. Boonstra, and J. M. Wojtowicz, "Hippocampal Neurogenesis in Food-Storing Red Squirrels: The Impact of Age and Spatial Behavior," *Genes, Brain, and Behavior* 9 (2010): 583–591.

13. Ibid.; D. F. Sherry, L. F. Jacobs, and S. J. C. Gaulin, "Spatial Memory and Adaptive Specialization of the Hippocampus," *Trends in Neurosciences* 15 (1992): 298–303.

14. H. J. Jerison, "Brain Size and the Evolution of Mind," James Arthur Lecture on the Evolution of the Human Brain, American Museum of Natural History, New York, 1991.

15. J. A. Amat et al., "Correlates of Intellectual Ability with Morphology of the Hippocampus and Amygdala in Healthy Adults," *Brain and Cognition* 66 (2008): 105–114.

16. E. A. Maguire et al., "Navigation-Related Structural Change in the Hippocampi of Taxi Drivers," *Proceedings of the National Academy of Sciences* 97 (2000): 4398–4403; E. A. Maguire, K. Woollett, and H. J. Spiers, "London Taxi Drivers and Bus Drivers: A Structural MRI Neuropsychological Analysis," *Hippocampus* 16 (2006): 1091–1101.

17. K. Woollett, J. Glensman, and E. A. Maguire, "Non-Spatial Expertise and Hippocampal Gray Matter Volume in Humans," *Hippocampus* 18 (2008): 981–984.

18. T. L. Davidson et al., "A Potential Role for the Hippocampus in Energy Intake and Body Weight Regulation," *Current Opinion in Pharmacology* 7 (2007): 613–616.

19. N. Germain et al., "Constitutional Thinness and Lean Anorexia Nervosa Display Opposite Concentrations of Peptide YY, Glucagon-Like Peptide 1, Ghrelin, and Leptin," *American Journal of Clinical Nutrition* 85 (2007): 967–971.

20. S. A. Farr, W. A. Banks, and J. E. Morley, "Effects of Leptin on Memory Processing," *Peptides* 27 (2006): 1420–1425; J. Harvey, N. Solovyova, and A. Irving, "Leptin and Its Role in Hippocampal Synaptic Plasticity," *Progress in Lipid Research* 45 (2006): 369–378; P. R. Moult and J. Harvey, "Hormonal Regulation of Hippocampal Dendritic Morphology and Synaptic Plasticity," *Cell Adhesion and Migration* 2 (2008): 269–275.

21. P. K. Olszewski, H. B. Schiöth, and A. S. Levine, "Ghrelin in the CNS: From Hunger to a Rewarding and Memorable Meal?" *Brain Research Reviews* 58 (2008): 160–170.

22. Davidson et al., "Potential Role for the Hippocampus."

23. C. Messier, "Glucose Improvement of Memory: A Review," *European Journal of Pharmacology* 490 (2004): 33–57.

24. Ibid.

25. A. L. Macready et al., "Flavonoids and Cognitive Function: A Review of Human Randomized Controlled Trial Studies and Recommendations for Future Studies," *Genes and Nutrition* 4 (2009): 227–243; J. P. E. Spencer, "The Impact of Fruit Flavonoids on Memory and Cognition," *British Journal of Nutrition* 104 (2010): S40–S47.

26. Spencer, "Impact of Fruit Flavonoids."

27. G. W. Arendash and C. Cao, "Caffeine and Coffee as Therapeutics against Alzheimer's Disease," *Journal of Alzheimer's Disease* 20 (2010): S117–S126.

28. P. Wostyn et al., "Increased Cerebrospinal Fluid Production as a Possible Mechanism Underlying Caffeine's Protective Effect against Alzheimer's Disease," *International Journal of Alzheimer's Disease* 2011 (2011): 617420.

29. Arendash and Cao, "Caffeine and Coffee."

30. W. Grimes, "First, a Little Something from the Chef . . . Very, Very Little," *New York Times*, July 22, 1998.

31. D. R. Paul et al., "Validation of a Food Frequency Questionnaire by Direct Measurement of Habitual Ad Libitum Food Intake," *American Journal of Epidemiology* 162 (2005): 806–814; A. F. Subar et al., "Comparative Validation of the Block, Willett, and National Cancer Institute Food

Frequency Questionnaires," *American Journal of Epidemiology* 154 (2001): 1089–1099.

32. Paul et al., "Validation of a Food Frequency Questionnaire"; Subar et al., "Comparative Validation"; W. Willett, "A Further Look at Dietary Questionnaire Validation," *American Journal of Epidemiology* 154 (2001): 1100–1102; G. Block, "Another Perspective on Food Frequency Question-naires," *American Journal of Epidemiology* 154 (2001): 1103–1104.

33. Paul et al., "Validation of a Food Frequency Questionnaire," 812.

34. Willett, "A Further Look at Dietary Questionnaire Validation," 1100.

35. B. Wansink, *Mindless Eating: Why We Eat More Than We Think* (New York: Bantam, 2006).

36. Ibid., 40.

37. P. Rozin et al., "What Causes Humans to Begin and End a Meal? A Role for Memory for What Has Been Eaten, as Evidenced by a Study of Multiple Meal Eating in Amnesic Patients," *Psychological Science* 9 (1998): 392–396.

38. Ibid., 394.

39. S. Higgs et al., "Sensory-Specific Satiety Is Intact in Amnesics Who Eat Multiple Meals," *Psychological Science* 19 (2008): 623–628.

40. I. L. Bernstein, "Food Aversion Learning: A Risk Factor of Nutri-tional Problems in the Elderly," *Physiology and Behavior* 66 (1999): 199–201; C. C. Horn, "Why Is the Neurobiology of Nausea and Vomiting So Im-portant?" *Appetite* 50 (2008): 430–434.

41. P. Rozin, "Psychobiological Perspectives on Food Preferences and Avoidances," in *Food and Evolution: Toward a Theory of Human Food Habits*, ed. M. Harris and E. B. Ross, 181–205 (Philadelphia: Temple University Press, 1987).

42. F. Bermúdez-Rattoni, "Molecular Mechanisms of Taste-Recognition Memory," *Nature Reviews Neuroscience* 5 (2004): 209–217.

43. Ibid.

44. K. Koops, W. C. McGrew, and T. Matsuzawa, "Do Chimpanzees *(Pan troglodytes)* Use Cleavers and Anvils to Fracture *Treculia africana* Fruits? Preliminary Data on a New Form of Percussive Technology," *Pri-mates* 51 (2010): 175–178; W. C. McGrew, "Primatology: Advanced Ape Technology," *Current Biology* 14 (2004): R1046–R1047; D. J. Povinelli, J. E. Reaux, and S. H. Frey, "Chimpanzees' Context-Dependent Tool Use Pro-vides Evidence for Separable Representations of Hand and Tool Even

during Active Use within Peripersonal Space," *Neuropsychologia* 48 (2010): 243–247.

45. A. D. Baddeley, "Is Working Memory Still Working?" *American Psychologist* 56 (2001): 851–864.

46. F. L. Coolidge and T. Wynn, "Working Memory, Its Executive Functions, and the Emergence of Modern Thinking," *Cambridge Archaeological Journal* 15 (2005): 5–26; T. Wynn and F. L. Coolidge, "Beyond Symbolism and Language: An Introduction to Supplement 1, *Working Memory*," *Current Anthropology* 51 (2010): S5–S16. The concept of the "modern mind" should be taken with a grain of salt, because it presupposes an ability to precisely define modern as opposed to premodern, and to sharply demarcate those hominins who possessed a modern mind from those who did not. See J. J. Shea, "*Homo sapiens* Is as *Homo sapiens* Was," *Current Anthropology* 52 (2011): 1–35.

47. D. E. J. Linden, "The Working Memory Networks of the Human Brain," *Neuroscientist* 13 (2007): 257–267; D. M. Barch and E. Smith, "The Cognitive Neuroscience of Working Memory: Relevance to CNTRICS and Schizophrenia," *Biological Psychiatry* 64 (2008): 11–17; T. Klingberg, "Training and Plasticity of Working Memory," *Trends in Cognitive Sciences* 14 (2010): 317–324.

48. C. P. Beaman, "Working Memory and Working Attention," *Current Anthropology* 51 (2010): S27–S38; M. N. Haidle, "Working-Memory Capacity and the Evolution of Modern Cognitive Potential," *Current Anthropology* 51 (2010):S149–S166.

49. Shea, "*Homo sapiens* Is as *Homo sapiens* Was."

50. G. O. Einstein et al., "Multiple Processes in Prospective Memory Retrieval: Factors Determining Monitoring Versus Spontaneous Retrieval," *Journal of Experimental Psychology: General* 134 (2005): 327–342; J. Fish, B. A. Wilson, and T. Manly, "The Assessment and Rehabilitation of Prospective Memory Problems in People with Neurological Disorders: A Review," *Neuropsychological Rehabilitation* 20 (2010): 161–179.

51. Fish, Wilson, and Manly, "Assessment and Rehabilitation"; P. W. Burgess, "Strategy Application Disorder: The Role of the Frontal Lobes in Human Multitasking," *Psychological Research* 63 (2000): 279–288; P. W. Burgess, A. Quayle, and C. D. Frith, "Brain Regions Involved in Prospective Memory as Determined by Positron Emission Tomography," *Neuropsychologia* 39 (2001): 545; H. E. M. den Ouden et al., "Thinking about Intentions,"

NeuroImage 28 (2005): 787–796; Y. Wang et al., "Meta-Analysis of Prospective Memory in Schizophrenia: Nature, Extent, and Correlates," *Schizophrenia Research* 114 (2009): 64–70.

52. R. Wrangham, *Catching Fire: How Cooking Made Us Human* (New York: Basic Books, 2009).

53. D. E. Sutton, *Remembrance of Repasts* (Oxford: Berg, 2001); D. E. Sutton, "A Tale of Easter Ovens: Food and Collective Memory," *Social Research* 75 (2008): 157–180.

54. Sutton, *Remembrance of Repasts*, 28.

55. Ibid., 29.

56. T. Suddendorf, "Episodic Memory versus Episodic Foresight: Similarities and Differences," *Wiley Interdisciplinary Reviews Cognitive Sciences* 1 (2009): 99–107; T. Suddendorf and M. C. Corballis, "The Evolution of Foresight: What Is Mental Time Travel, and Is It Unique to Humans?" *Behavioral and Brain Sciences* 30 (2007): 299–351; T. Suddendorf, D. R. Addis, and M. C. Corballis, "Mental Time Travel and the Shaping of the Human Mind," *Philosophical Transactions of the Royal Society B* 364 (2009): 1317–1324.

57. Suddendorf, "Episodic Memory."

58. J. D. Holtzman, "Food and Memory," *Annual Review of Anthropology* 35 (2006): 361–378.

59. J. Siskind, "The Invention of Thanksgiving: A Ritual of American Nationality," *Critique of Anthropology* 12 (1992): 167–191, quote from 185.

60. M. Halbwachs, *The Collective Memory* (New York: Harper and Row Colophon, 1980), 44.

6. Categories: Good Food, Bad Food, Yes Food, No Food

1. A. Escoffier, *Memories of My Life*, trans. L. Escoffier (New York: Van Nostrand Reinhold, 1997), 33.

2. See, for example, M. Douglas, *Purity and Danger: An Analysis of Concepts of Pollution and Taboo* (New York: Praeger, 1966); M. Harris, *Good to Eat: Riddles of Food and Culture* (Prospect Heights, IL: Waveland, 1985).

3. R. J. Sullivan and E. H. Hagen, "Psychotropic Substance-Seeking: Evolutionary Pathology or Adaptation?" *Addiction* 97 (2002): 389–400.

4. J. Waugh, "DNA Barcoding in Animal Species: Progress, Potential, and Pitfalls," *BioEssays* 29 (2007): 188–197; see also the PhyloCode Project at www.ohio.edu/phylocode/index.html.

5. S. Atran, "Folk Biology and the Anthropology of Science: Cognitive Universals and Cultural Particulars," *Behavioral and Brain Sciences* 21 (1998): 547–609; M. Bang, D. L. Medin, and S. Atran, "Cultural Mosaics and Mental Models of Nature," *Proceedings of the National Academy of Sciences* 104 (2007): 13868–13874.

6. Atran, "Folk Biology and the Anthropology of Science."

7. R. Bulmer, "Why Is the Cassowary Not a Bird? A Problem of Zoological Taxonomy among the Karam of the New Guinea Highlands," *Man* 2 (n.s.) (1967): 5–25; I. S. Majnep and R. N. H. Bulmer, *Bird of My Kalam Country (Mnmon yad Kalam Yakt)* (Auckland: Auckland University Press, 1977).

8. Bulmer, "Why Is the Cassowary Not a Bird?"

9. Ibid., 17.

10. S. R. Kellert, "The Biological Basis for Human Values of Nature," in *The Biophilia Hypothesis*, ed. S. R. Kellert and E. O. Wilson, 42–72 (Washington, DC: Island Press, 1993).

11. F. G. Ashby et al., "A Neuropsychological Theory of Multiple Systems of Category Learning," *Psychological Review* 105 (1998): 442–481; F. G. Ashby and W. T. Maddox, "Human Category Learning," *Annual Review of Psychology* 56 (2005): 149–178; B. Z. Mahon and A. Caramazza, "Concepts and Categories: A Cognitive Neuropsychological Perspective," *Annual Review of Psychology* 60 (2009): 27–51.

12. Ashby and Maddox, "Human Category Learning."

13. Ibid., 167.

14. C. Piras, *Culinaria Italy: Pasta, Pesto, Passion* (Potsdam: H. F. Ullmann, 2007).

15. Ashby et al., "A Neuropsychological Theory."

16. H. McGee, *On Food and Cooking: The Science and Lore of the Kitchen* (New York: Scribner, 2004), 153.

17. Ashby and Maddox, "Human Category Learning," 169.

18. C. A. Seger and E. K. Miller, "Category Learning in the Brain," *Annual Review of Neuroscience* 33 (2010): 203–219, quote from 213.

19. P. Fusar-Poli et al., "Functional Atlas of Emotional Faces Processing: A Voxel-Based Meta-Analysis of 105 Functional Magnetic Resonance Imaging Studies," *Journal of Psychiatry and Neuroscience* 34 (2009): 418–432.

20. J. S. Foer, *Eating Animals* (New York: Back Bay Books, 2009), 6.

21. G. A. Miller and P. M. Gildea, "How Children Learn Words," in *The Emergence of Language: Development and Evolution*, ed. W. S.-Y. Wang, 150–158 (New York: W. H. Freeman, 1991).

22. Oxford Dictionaries Online, "How Many Words Are There in the English Language?" 2010, www.oxforddictionaries.com/page/93.

23. P. T. Schoenemann, "Syntax as an Emergent Characteristic of the Evolution of Semantic Complexity," *Minds and Machines* 9 (1999): 309–346.

24. S. Savage-Rumbaugh and D. Rumbaugh, "The Emergence of Language," in *Tools, Language, and Cognition in Human Evolution*, ed. K. R. Gibson and T. Ingold, 86–108 (Cambridge: Cambridge University Press, 1993); S. Savage-Rumbaugh, S. G. Shanker, and T. J. Taylor, *Apes, Language, and the Human Mind* (New York: Oxford University Press, 1998); P. T. Schoenemann, "Conceptual Complexity and the Brain: Understanding Language Origins," in *Language Acquisition, Change, and Emergence: Essays in Evolutionary Linguistics*, ed. W. S.-Y. Wang and J. W. Minett, 47–94 (Hong Kong: City University of Hong Kong Press, 2005).

25. For a discussion of theories of language origins and the brain, see J. S. Allen, *The Lives of the Brain: Human Evolution and the Organ of Mind* (Cambridge, MA: Belknap Press, 2009), 232–272.

26. J. Painter, J.-H. Rah, and Y.-K. Lee, "Comparison of International Food Guide Pictorial Representations," *Journal of the American Dietetic Association* 102 (2002): 483–489, quote from 489.

27. M. Nestle, *Food Politics: How the Food Industry Influences Nutrition and Health*, rev. ed. (Berkeley: University of California Press, 2007).

28. Ibid., 27; S. P. Murphy and S. I. Barr, "Food Guides Reflect Similarities and Differences in Dietary Guidance in Three Countries (Japan, Canada, and the United States)," *Nutrition Reviews* 65 (2007): 141–148.

29. S. W. Katamay et al., *"Eating Well with Canada's Food Guide (2007):* Development of the Food Intake Pattern," *Nutrition Reviews* 65 (2007): 155–166; N. Yoshiike et al., "A New Food Guide in Japan: *The Japanese Food Guide Spinning Top*," *Nutrition Reviews* 65 (2007): 149–154.

30. M. Pollan, *In Defense of Food: An Eater's Manifesto* (New York: Penguin, 2008).

31. G. Taubes, *Good Calories, Bad Calories* (New York: Anchor Books, 2007).

32. Ibid., 28.

33. C. D. Naylor and J. M. Paterson, "Cholesterol Policy and the Primary Prevention of Coronary Disease: Reflections on Clinical and Population Strategies," *Annual Review of Nutrition* 16 (1996): 349–382.

34. A. M. Brownawell and M. C. Falk, "Cholesterol: Where Science and Public Health Policy Intersect," *Nutrition Reviews* 68 (2010): 355–364.

35. Ibid.

36. Ibid., 361.

37. Naylor and Paterson, "Cholesterol Policy and the Primary Prevention of Coronary Disease."

38. Taubes, *Good Calories, Bad Calories.* See page 19.

39. J. Haidt et al., "Body, Psyche, and Culture: The Relationship between Disgust and Morality," *Psychology and Developing Societies* 9 (1997): 107–131, quote from 121.

40. A. R. Damasio, *Descartes' Error: Emotion, Reason, and the Human Brain* (New York: Avon, 1994).

41. Ibid., 173.

42. C. M. Funk and M. S. Gazzaniga, "The Functional Brain Architecture of Human Morality," *Current Opinion in Neurobiology* 19 (2009): 678–681.

43. T. Wheatley and J. Haidt, "Hypnotic Disgust Makes Moral Judgments More Severe," *Psychological Science* 16 (2005): 780–784.

44. J. S. Borg, D. Lieberman, and K. A. Kiehl, "Infection, Incest, and Iniquity: Investigating the Neural Correlates of Disgust and Morality," *Journal of Cognitive Neuroscience* 20 (2008): 1529–1546.

45. J. Wechsberg, *Blue Trout and Black Truffles* (New York: Alfred A. Knopf, 1954).

7. Food and the Creative Journey

1. S. Kawamura, "The Process of Sub-Culture Propagation among Japanese Macaques," *Primates* 2 (1959): 43–54.

2. T. Keller, S. Heller, and M. Ruhlman, *The French Laundry Cookbook* (New York: Artisan, 1999). Quote from p. 3.

3. F. Adrià, J. Soler, and A. Adrià, *A Day at El Bulli: An Insight into the Ideas, Methods, and Creativity of Ferran Adrià* (London: Phaidon, 2008).

4. Ibid., insert between 240 and 241.

5. G. Achatz, *Alinea* (Berkeley, CA: Ten Speed Press, 2008); G. Achatz, "Diner's Journal: What Grant Achatz Saw at El Bulli," *New York Times*, February 16, 2010.

6. G. Cochran and H. Harpending, *The 10,000 Year Explosion* (New York: Basic Books, 2009), 127.

7. D. K. Simonton, *Origins of Genius: Darwinian Perspectives on Creativity* (New York: Oxford University Press, 1999).

8. A. Flaherty, "Frontotemporal and Dopaminergic Control of Idea Generation and Creative Drive," *Journal of Comparative Neurology* 493 (2005): 147–153, quote from 147.

9. G. Miller, *The Mating Mind* (New York: Anchor Books, 2000).

10. S. Blackmore, *The Meme Machine* (Oxford: Oxford University Press, 1999).

11. C. Stanford, J. S. Allen, and S. C. Antón, *Biological Anthropology: The Natural History of Humankind*, 2nd ed. (Upper Saddle River, NJ: Prentice-Hall, 2009).

12. Discussed in T. I. Lubart, "Models of the Creative Process: Past, Present, and Future," *Creativity Research Journal* 13 (2000–2001): 295–308.

13. Ibid.

14. V. Drago et al., "What's Inside the Art? The Influence of Fronto-temporal Dementia in Art Production," *Neurology* 67 (2006): 1285–1287.

15. L. C. de Souza et al., "Poor Creativity in Frontotemporal Dementia: A Window into the Neural Basis of the Creative Mind," *Neuropsychologia* 48 (2010): 3733–3742.

16. Ibid.

17. Flaherty, "Frontotemporal and Dopaminergic Control."

18. R. E. Jung et al., "Neuroanatomy of Creativity," *Human Brain Mapping* 31 (2010): 398–409. The Creative Achievement Questionnaire is designed to assess creativity in ten different domains (visual arts, music, etc.); Jung and colleagues also measured "divergent thinking" (another experimental proxy for creativity) using a variety of design tasks, with the results consensually assessed by raters into a "composite creativity index." Magnetic resonance images of the subjects' brains (there were sixty-one subjects in total) were compared to one another, and a computer program was used to measure the correlation between the various creative measures and the cortical thickness—the surface gray matter—of the subjects' brains.

19. H. Takeuchi et al., "Regional Gray Matter Volume of Dopaminergic System Associate with Creativity: Evidence from Voxel-Based Morphometry," *NeuroImage* 51 (2010): 578–585.

20. O. de Manzano et al., "Thinking Outside a Less Intact Box: Thalamic Dopamine D2 Receptor Densities Are Negatively Related to Psychometric Creativity in Healthy Individuals," *PLoS One* 5 (2010): e10670.

21. Flaherty, "Frontotemporal and Dopaminergic Control."

22. A. Harrington, *Medicine, Mind, and the Double Brain* (Princeton: Princeton University Press, 1987); S. Finger, *Minds Behind the Brain* (New York: Oxford University Press, 2000).

23. R. Sperry, "Roger W. Sperry—Nobel Lecture," 1981. Available at Nobelprize.org, http://nobelprize.org/nobel_prizes/medicine/laureates/1981/sperry-lecture.html.

24. M. Jung-Beeman et al., "Neural Activity When People Solve Verbal Problems with Insight," *PLoS Biology* 2 (2004): 0500–0510.

25. A. Dietrich and R. Kanso, "A Review of EEG, ERP, and Neuroimaging Studies of Creativity and Insight," *Psychological Bulletin* 136 (2010): 822–848; R. D. Whitman, E. Holcomb, and J. Zanes, "Hemispheric Collaboration in Creative Subjects: Cross-Hemisphere Priming in a Lexical Decision Task," *Creativity Research Journal* 22 (2010): 109–118.

26. K. M. Mihov, M. Denzler, and J. Förster, "Hemispheric Specialization and Creative Thinking: A Meta-Analytic Review of Lateralization of Creativity," *Brain and Cognition* 72 (2010): 442–448.

27. R. E. Jung et al., "White Matter Integrity, Creativity, and Psychopathology: Disentangling Constructs with Diffusion Tensor Imaging," *PLoS One* 5 (2010): e9818.

28. S. T. Hunter, K. E. Bedell, and M. D. Mumford, "Climate for Creativity: A Quantitative Review," *Creativity Research Journal* 19 (2007): 69–90.

29. Ibid.

30. Ibid.

31. V. Chossat and O. Gergaud, "Expert Opinion and Gastronomy: The Recipe for Success," *Journal of Cultural Economics* 27 (2003): 127–141.

32. Adrià, Soler, and Adrià, *A Day at El Bulli*; Achatz, *Alinea*; Keller, Heller, and Ruhlman, *The French Laundry Cookbook*.

33. J.-S. Horng and M.-L. Hu, "The Mystery in the Kitchen: Culinary Creativity," *Creativity Research Journal* 20 (2008): 221–230; J.-S. Horng and

M.-L. Hu, "The Creative Culinary Process: Constructing and Extending a Four-Component Model," *Creativity Research Journal* 21 (2009): 376–383.

34. See www.foodandwine.com/best_new_chefs/by_name[0].

35. L. Heldke, "Let's Cook Thai: Recipes for Colonialism," in *Food and Culture: A Reader*, ed. C. Counihan and P. van Esterik, 2nd ed., 327–341 (New York: Routledge, 2008).

36. Ibid., 334.

37. J. Anderson, *The American Century Cookbook* (New York: Clarkson Potter, 1997), 3.

38. Amana Heritage Society, *Guten Appetit from Amana Kitchens* (Amana, IA: Amana Preservation Foundation, 1985).

39. J. Baer and J. C. Kaufman, "Gender Differences in Creativity," *Journal of Creative Behavior* 42 (2008): 75–105, quote from 98.

8. THEORY OF MIND, THEORY OF FOOD?

1. K. L. Sakai, "Language Acquisition and Brain Development," *Science* 310 (2005): 815–819.

2. D. Premack and G. Woodruff, "Does the Chimpanzee Have a Theory of Mind?" *Behavioral and Brain Sciences* 4 (1978): 515–526; D. Premack and G. Woodruff, "Chimpanzee Problem-Solving: A Test for Comprehension," *Science* 202 (1978): 532–535.

3. A. M. Leslie, " 'Theory of Mind' as a Mechanism of Selective Attention," in *The New Cognitive Neurosciences*, ed. M. S. Gazzaniga, 2nd ed., 1235–1247 (Cambridge, MA: MIT Press, 2000).

4. Ibid., 1235.

5. S. Baron-Cohen, "The Cognitive Neuroscience of Autism: Evolutionary Approaches," in *The New Cognitive Neurosciences*, ed. M. S. Gazzaniga, 2nd ed., 1249–1257 (Cambridge, MA: MIT Press, 2000); G. J. Pickup, "Relationship between Theory of Mind and Executive Function in Schizophrenia: A Systematic Review," *Psychopathology* 41 (2008): 206–213.

6. S. Baron-Cohen, "Autism: The Empathizing-Systemizing (E-S) Theory," *Annals of the New York Academy of Sciences* 1156 (2009): 68–80, quote from 69.

7. S. J. Carrington and A. J. Bailey, "Are There Theory of Mind Regions in the Brain? A Review of the Neuroimaging Literature," *Human Brain Mapping* 30 (2008): 2313–2335.

8. M. A. Just and S. Varma, "The Organization of Thinking: What Functional Brain Imaging Reveals about the Neuroarchitecture of Complex Cognition," *Cognitive, Affective, and Behavioral Neuroscience* 7 (2007): 153–191.

9. Ibid., 154.

10. M. Tomasello and J. Call, *Primate Cognition* (New York: Oxford University Press, 1997).

11. J. Call and M. Tomasello, "Does the Chimpanzee Have a Theory of Mind? 30 Years Later," *Trends in Cognitive Sciences* 12 (2008): 187–192.

12. Ibid., 131.

13. S. L. Anzman, B. Y. Rollins, and L. L. Birch, "Parental Influence on Children's Early Eating Environments and Obesity Risk: Implications for Prevention," *International Journal of Obesity* 34 (2010): 1116–1124.

14. S. L. Johnston, "Evolutionary Dimensions of Human Meal Patterns," *American Journal of Human Biology* 23 (2011): 262–263.

15. K. R. Daffner, "Promoting Successful Cognitive Aging: A Comprehensive Review," *Journal of Alzheimer's Disease* 19 (2010): 1101–1022; B. R. Reed et al., "Cognitive Activities during Adulthood Are More Important than Education in Building Reserve," *Journal of the International Neuropsychological Society* 17 (2011): 615–624.

16. M. K. Rohr and F. R. Lang, "Aging Well Together—A Mini-Review," *Gerontology* 55 (2009): 333–343; B. D. James et al., "Late-Life Social Activity and Cognitive Decline in Old Age," *Journal of the International Neuropsychological Society* 17 (2011): 998–1005.

ACKNOWLEDGMENTS

Many of the ideas expressed in this book have been baking (or half-baking) slowly over many years, and under the influence of many friends and colleagues. In graduate school, much of my thinking on food, diet, and evolution was strongly shaped by Fatimah Jackson, and especially, by Katie Milton, whose work applying what we know about nonhuman primate diets toward a greater understanding of human evolution has been foundational. Since then, I have spent many hours discussing food and human biology with Alex Brewis-Slade, food and archaeology with Peter Sheppard, and food and the biocultural nature of the human condition with Andrea Wiley. This book would not have been the same without those discussions. My textbook-writing colleagues Craig Stanford and Susan Antón willingly answered any of my questions in their areas of expertise. My two former graduate students Susan Cheer (milk/lactose) and Roger Sullivan (betel nut) taught me quite a bit about the nature of human ingestion and digestion.

I thank Andrea Wiley and Peter Sheppard for reading all or part of the manuscript, as well as two anonymous reviewers, and for their comments, which improved the final product immeasurably. Linda Blackford was also kind enough to read the manuscript and provide me with her take on it as a food enthusiast and a journalist. Joel Bruss did the illustrations, and I thank him for quite literally donating his own brain to the project. I thank Ricardo Cardenas for his insights into the restaurant business and Mexican cuisine.

This book would not have been possible without the leadership and support of Michael Fisher at Harvard University Press. He has been involved with the project since its inception, and has provided invaluable guidance through each stage of the writing and review process. I also thank Lauren Esdaile at HUP for handling the production tasks of the book. Thanks very much also to Sue Warga for the excellent and efficient copyediting.

My bosses and colleagues Hanna and Antonio Damasio have long provided me with a stimulating intellectual home, first at the University of Iowa, and currently at the Dornsife Cognitive Neuroscience Imaging Center and the Brain and Creativity Institute at the University of Southern California. I cannot thank them enough for their support over the years, which has allowed me to pursue neurocognitive and evolutionary research concerning a wide range of topics.

Finally, I thank my wife, Stephanie Sheffield, for reading the chapters as they emerged, for her comments on them, and for her love and support and gardening skills. I also thank my sons, Reid and Perry, for being guardedly enthusiastic about much of what I cook for them.

INDEX